World Regional Geography

World Regional Geography

CAITLIN FINLAYSON

World Regional Geography by Caitlin Finlayson is licensed under a Creative Commons Attribution-NonCommercial-ShareAlike 4.0 International License, except where otherwise noted.

This version of the text was last updated on 8/9/2022

Contents

Preface 1

PART I. **CHAPTERS**

1. Introduction 3
   1.1 The Where and the Why 3
   1.2 The Spatial Perspective 5
   1.3 Core and Periphery 8
   1.4 The Physical Setting 9
   1.5 The Human Setting 13
   1.6 The World's Regions 17
   1.7 Sub-disciplines of Geography 21
   1.8 Globalization and Inequality 23

2. Europe 26
   2.1 European Physical Geography and Boundaries 26
   2.2 Cooperation and Control in Europe 29
   2.3 The Industrial Revolution 35
   2.4 European Migration 39
   2.5 Shifting National Identities 40
   2.6 Current Migration Patterns and Debates 43

3. Russia 46
   3.1 Russia's Physical Geography and Climate 46
   3.2 Settlement and Development Challenges 50
   3.3 Russian History and Expansion 53
   3.4 Russian Multiculturalism and Tension 55
   3.5 Economics and Development in the Soviet Union 61
   3.6 The Modern Russian Landscape 63

4. North America . . . 66
   4.1 North America's Physical Setting . . . 66
   4.2 North American History and Settlement . . . 71
   4.3 Industrial Development in North America . . . 75
   4.4 The North American Urban Landscape . . . 77
   4.5 Patterns of Inequality in North America . . . 82
   4.6 North America's Global Connections . . . 84

5. Middle and South America . . . 86
   5.1 The Geographic Features of Middle and South America . . . 86
   5.2 Colonization and Conquest in Middle America . . . 93
   5.3 The South American Colonial Landscape . . . 97
   5.4 Urban Development in South America . . . 100
   5.5 Income Inequality in Middle and South America . . . 104
   5.6 Patterns of Globalization in Middle and South America . . . 106

6. Sub-Saharan Africa . . . 109
   6.1 The Physical Landscape of Sub-Saharan Africa . . . 109
   6.2 Pre-Colonial Sub-Saharan Africa . . . 114
   6.3 Sub-Saharan African Colonization . . . 117
   6.4 The Modern Sub-Saharan African Landscape . . . 119
   6.5 Economics and Globalization in Sub-Saharan Africa . . . 123

7. North Africa and Southwest Asia . . . 127
   North Africa and Southwest Asia's Key Geographic Features . . . 127
   7.2 Cultural Adaptations in North Africa and Southwest Asia . . . 132
   7.3 The Religious Hearths of North Africa and Southwest Asia . . . 136
   7.4 Conquest in North Africa and Southwest Asia . . . 140
   7.5 The Modern Political Landscape of North Africa and Southwest Asia . . . 141
   7.6 Religious Conflict in North Africa and Southwest Asia . . . 144

8. South Asia . . . 147
   8.1 South Asia's Physical Landscape . . . 147
   8.2 Patterns of Human Settlement in South Asia . . . 152
   8.3 Cultural Groups in South Asia . . . 156
   8.4 South Asia's Population Dynamics . . . 161
   8.5 Future Challenges and Opportunities in South Asia . . . 163

9.	East and Southeast Asia	165
	9.1 The Physical Landscape of East and Southeast Asia	165
	9.2 Natural Hazards in East and Southeast Asia	168
	9.3 East and Southeast Asia's History and Settlement	171
	9.4 Political Conflicts and Changes East and Southeast Asia	176
	9.5 Patterns of Economic Development in East and Southeast Asia	178
10.	Oceania	184
	10.1 The Physical Landscape of Oceania	184
	10.2 The World's Oceans and Polar Frontiers	189
	10.3 Biogeography in Australia and the Pacific	194
	10.4 The Patterns of Human Settlement in Australia and the Pacific	196
	10.5 The Changing Landscape of Oceania	200
	Glossary	205

Preface

Geography is a discipline of explorers. Some geographers explore the world using satellite imagery and others by interviewing members of an indigenous community in an isolated area. What unites geographers everywhere is a desire to dig deeper, a desire to better understand why the spatial patterns and unique features we find in the world exist and how they interact and change. World Regional Geography presents an overview of the discipline by introducing students to key themes and concepts in the discipline of geography through a study of the world's regions.

In a traditional World Regional Geography textbook, chapters are arranged around the various regions of the world with each chapter focusing on the geographic features of the particular region. Concepts such as climate, physical features, culture, economics, and politics are discussed in every chapter and particular places and names of physical features found in each region are emphasized. In essence, most World Regional Geography textbooks privilege breadth over depth.

There are two key problems with this traditional approach. First, most regional chapters follow the same basic outline of topics, perhaps beginning with physical features, then outlining historical developments, and then moving on to culture and economics. Countries and specific places within the region are emphasized rather than the patterns found across the region as a whole. There is rarely an over-arching theme or story that connects the regions to one another. Secondly, in most primary-level geography courses, breadth is already emphasized. Students may take map quizzes or learn a list of physical features, but have little exposure to the depth of concepts and theories that are central to geography as a discipline.

This book takes a different approach. Rather than present students with a broad, novice-level introduction to geography, emphasizing places and vocabulary terms, this text approaches geography as experts understand the discipline, focusing on connections and an in-depth understanding of core themes. This thematic approach, informed by pedagogical research, provides students with an introduction to thinking geographically. Instead of repeating the same several themes each chapter, this text emphasizes depth over breadth by arranging each chapter around a central theme and then exploring that theme in detail as it applies to the particular region. In addition, while chapters are designed to stand alone and be rearranged or eliminated at the instructor's discretion, the theme of globalization and inequality unites all of the regions discussed. This core focus enables students to draw connections between regions and to better understand the interconnectedness of our world. Furthermore, the focus on both globalization and inequality helps demonstrate the real-world application of the concepts discussed. Colonialism, for instance, rather than a historical relic, becomes a force that has shaped geography and informs social justice. This thematic approach is also intended to facilitate active learning and would be suitable for a flipped or team-based learning-style course since it more easily integrates case studies and higher-order thinking than the traditional model.

Each chapter begins with a list of learning objectives. This text was written with the backward course design model in mind and the content of each chapter was structured around these learning objectives. Because of this backward design focus, the length of each chapter is considerably shorter than most traditional textbooks. The intention is for the instructor to supplement the text with problems, case studies, and news articles and to use the text as a springboard for discussing deeper issues. The chapters are written in an accessible style, often addressing the student directly, and the author's voice has intentionally tried to remain present in the text. Following the Washington Post's gender-inclusive style guide, the singular they is intentionally used throughout the text. Rhetorical questions are also used to help students reflect on concepts and to encourage them to dig deeper and consider concepts from different perspectives.

Finally, a key difference between this text and others on the subject is that it is provided at no cost under the CC BY license. This means that the content can be distributed, remixed, tweaked, or built upon simply by crediting the author. Geography is an open discipline. In truth, anyone can be a geographer as long as they are curious about the world around them.

This isn't a perfect text and it doesn't attempt to be. In emphasizing depth over breadth, some content was sacrificed. However, the intention is that students will not only know the material much more deeply, but in doing so, will also develop a passion for geography and a geographical imagination that will continue beyond this course.

Happy exploring, geographers.

CHAPTER 1

Introduction

> **Learning Objectives**
>
> - Understand the principles of geographic study
> - Summarize the key physical and human features of the world
> - Distinguish between different types of regions
> - Understand the major subfields of geography and their key conceptual frameworks
> - Describe the process of globalization and the principal measures of inequality

1.1 THE WHERE AND THE WHY

What is "geography"? It might seem like a simple enough term to define. In middle school or high school, your answer might have been something to do with the study of maps, of where things were located in the world. In fact, much of primary and secondary school geography is explicitly focused on the where, answering questions like where a particular country is located, what a country's capital is, and where major landforms are located. Just as simple arithmetic operations form the backbone of mathematics as a discipline, these kinds of questions are foundational to geographic study. However, one wouldn't likely define math as the study of calculators or of multiplication tables. Similarly, there is much more to geography and geographic inquiry than the study of maps.

Geographers seek to answer both the "where" and the "why." Simply knowing where a country is located is certainly helpful, but geographers dig deeper: why is it located there? Why does it have a particular shape, and how does this shape affect how it interacts with its neighbors and its access to resources? Why do the people of the country have certain cultural features? Why does the country have a specific style of government? The list goes on and on, and as you might notice, incorporates a variety of historical, cultural, political, and physical features. This synthesis of the physical world and human activity is at the heart of the regional geographic approach.

The term "geography" comes from the Greek term *geo* meaning "the earth" and *graphia* meaning "to write," and many early geographers did exactly that: they wrote about the world. Ibn Battuta, for example, was a scholar from Morocco and traveled extensively across Africa and Asia in the 14th century CE. Eratosthenes is commonly considered to be the "Father of Geography," and in fact, he quite literally wrote the book on the subject in the third century BCE. His three-volume text, *Geographica*, included maps of the entire known world **(see Figure 1.1)**, including different climate zones, the locations of hundreds of different cities, and a coordinate system. This was a revolutionary and highly regarded text, especially for the time period. Eratosthenes is also credited as the first person to calculate the circumference of the Earth. Many early geographers, like Eratosthenes, were primarily cartographers, referring to people who scientifically study and create maps, and early maps, such as those used in Babylon, Polynesia, and the Arabian Peninsula, were often used for navigation. In the Middle Ages, as academic inquiry in Europe declined with the fall of the Roman Empire, Muslim geographer Muhammad al-Idrisi created one of the most advanced maps of pre-modern times, inspiring future geographers from the region.

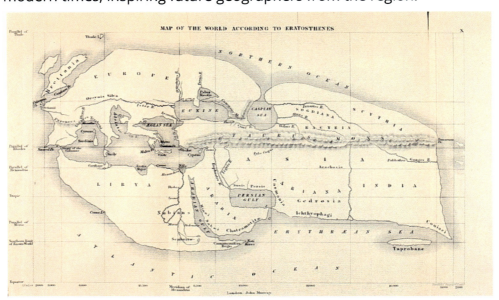

Figure 1.1: Reconstruction of Eratosthenes' Map of the Known World, c. 194 BCE (© E. H. Bunbury, A History of Ancient Geography among the Greeks and Romans from the Earliest Ages till the Fall of the Roman Empire, 1883, Public Domain)

Geography today, though using more advanced tools and techniques, draws on the foundations laid by these predecessors. What unites all geographers, whether they are travelers writing about the world's cultures or cartographers mapping new frontiers, is an attention to the spatial perspective. As geographer Harm deBlij once explained, there are three main ways to look at the world. One way is chronologically, as a historian might examine the sequence of world events. A second way is systematically, as a sociologist might explore the societal systems in place that help shape a given country's structures of inequality. The third way is spatially, and this is the geographic perspective. Geographers, when confronted with a global problem, immediately ask the questions "Where?" and "Why?" Although geography is a broad discipline that includes quantitative techniques like statistics and qualitative methods like interviews, all geographers share this common way of looking at the world from a spatial perspective.

1.2 THE SPATIAL PERSPECTIVE

At the heart of the spatial perspective is the question of "where," but there are a number of different ways to answer this question. **Relative location** refers to the location of a place relative to other places, and we commonly use relative location when giving directions to people. We might instruct them to turn "by the gas station on the corner," or say that we live "in the dorm across from the fountain." Another way to describe a place is by referring to its **absolute location**. Absolute location references an exact point on Earth and commonly uses specific coordinates like **latitude** and **longitude**. Lines of latitude and longitude are imaginary lines that circle the globe and form the geographic coordinate system (see **Figure 1.2**). Lines of latitude run laterally, parallel to the equator, and measure distances north or south of the equator. Lines of longitude, on the other hand, converge at the poles and measure distances east and west of the prime meridian.

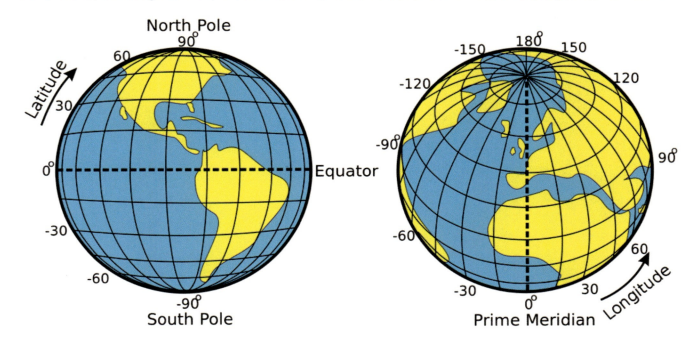

Figure 1.2: Lines of Latitude and Longitude (© Djexplo, Wikipedia Commons, CC0 1.0)

Every place on Earth has a precise location that can be measured with latitude and longitude. The location of the White House in Washington, DC, for example, is located at latitude 38.8977 °N and longitude 77.0365°W. Absolute location might also refer to details like elevation. The Dead Sea, located on the border of Jordan and Israel, is the lowest location on land, dipping down to 1,378 feet below sea level.

Historically, most maps were hand-drawn, but with the advent of computer technology came more advanced maps created with the aid of satellite technology. **Geographic information science** (GIS), sometimes also referred to as geographic information systems, uses computers and satellite imagery to capture, store, manipulate, analyze, manage, and present spatial data. GIS essentially uses layers of information and is often used to make decisions in a wide variety of contexts. An urban planner might use GIS to determine the best location for a new fire station, while a biologist might use GIS to map the migratory paths of birds. You might use GIS to get navigation directions from one place to another, layering place names, buildings, and roads.

One difficulty with map-making, even when using advanced technology, is that the earth is roughly a sphere while maps are generally flat. When converting the spherical Earth to a flat map, some **distortion** always occurs. A map projection , or a representation of Earth's surface on a flat plane, always distorts at least one of these four properties: area, shape, distance, and direction. Some maps preserve three of these properties, while significantly distorting another, while other maps seek to minimize overall distortion but distort each property somewhat. So which map projection is best? That depends on the purpose of the map. The Mercator projection , while significantly distorting the size of places near the poles, preserves angles and shapes, making it ideal for navigation (see **Figure 1.3**).

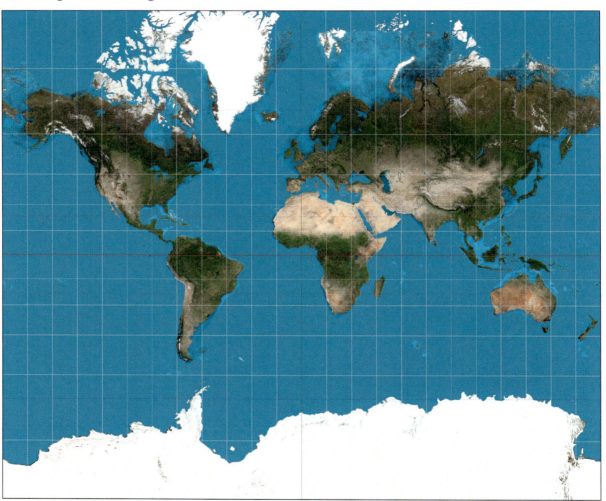

Figure 1.3: Mercator Projection (© Daniel R. Strebe, Wikimedia Commons, CC BY-SA 3.0)

The Winkel tripel projection is so-named because its creator, Oswald Winkel, sought to minimize three kinds of distortion: area, direction, and distance (see **Figure 1.4**). It has been used by the National Geographic Society since 1998 as the standard projection of world maps.

Figure 1.4: Winkel Tripel Projection (© Daniel R. Strebe, Wikimedia Commons, CC BY-SA 3.0)

When representing the Earth on a manageable-sized map, the actual size of location is reduced. **Scale** is the ratio between the distance between two locations on a map and the corresponding distance on Earth's surface. A 1:1000 scale map, for example, would mean that 1 meters on the map equals 1000 meters, or 1 kilometer, on Earth's surface. Scale can sometimes be a confusing concept for students, so it's important to remember that it refers to a ratio. It doesn't refer to the size of the map itself, but rather, how zoomed in or out the map is. A 1:1 scale map of your room would be the exact same size of your room – plenty of room for significant detail, but hard to fit into your glove compartment. As with map projections, the "best" scale for a map depends on what it's used for. If you're going on a walking tour of a historic town, a 1:5,000 scale map is commonly used. If you're a geography student looking at a map of the entire world, a 1:50,000,000 scale map would be appropriate. "Large" scale and "small" scale refer to the ratio, not to the size of the landmass on the map. 1 divided by 5,000 is 0.0002, which is a larger number than 1 divided by 50,000,000 (which is 0.00000002). Thus, a 1:5,000 scale map is considered "large" scale while 1:50,000,000 is considered "small" scale.

All maps have a purpose, whether it's to guide sailing ships, help students create a more accurate mental map of the world, or tell a story. The map projection, color scheme, scale, and labels are all decisions made by the mapmaker. Some argued that the widespread use of the Mercator projection, which made Africa look smaller relative to North America and Eurasia, led people to minimize the importance of Africa's political and economic issues. Just as texts can be critiqued for their style, message, and purpose, so too can maps be critiqued for the information and message they present.

The spatial perspective, and answering the question of "where," encompasses more than just static locations on a map. Often, answering the question of "where" relates to movement across space. **Diffusion** refers to the spreading of something from one place to another, and might relate to the physical movement of people or the spread of disease, or the diffusion of ideas, technology, or other intangible phenomena. Diffusion occurs for different reasons and at different rates. Just as static features of culture and the physical landscape can be mapped, geographers can also map the spread of various characteristics or ideas to study how they interact and change.

1.3 CORE AND PERIPHERY

One way of considering the location of places relative to one another is by examining their spatial interaction. In a given region, there is generally a core area, sometimes known as the central business district (CBD) and a hinterland, a German term literally meaning "the land behind" (see **Figure 1.5**). The hinterland is more sparsely populated than the core and is often where goods sold in the core are manufactured. It might include rural farmland, for example.

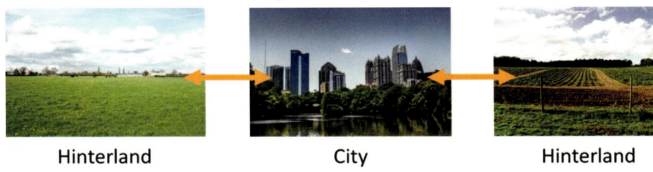

Figure 1.5: The Core and the Hinterland (Figure by author, Images courtesy of Espresso Addict, Wikimedia Commons; Mike – Flickr; Pam Brophy – Wikimedia Commons; CC BY-SA)

The core, on the other hand, is the commercial focus for the area where most goods and services are exchanged. The hinterland relies on the central city to sell its goods, but similarly the city relies on the hinterland to produce raw materials. Consider where the hinterland is located around your closest city; the hinterland is characteristically rural, while the core is urban. All countries contain core areas and hinterlands.

Globally, we can apply the hinterland-city model to an understanding of a global core and a global periphery (see **Figure 1.6**). The core areas are places of dominance, and these areas exert control over the surrounding periphery. Core areas are typically more developed and industrialized whereas the periphery is more rural and generally less developed. Unlike the interactions between the city and the hinterland, economic exchange between the core and periphery is characteristically one-sided, creating wealth for the core and patterns of uneven development. However, these interactions do contribute to economic stability in the periphery. Some argue that it benefits the core countries to keep the periphery peripheral; in other words, if the periphery can remain underdeveloped, they are more likely to sell cheap goods to the core. This generates more wealth for core areas and contributes to their continued influence and economic strength.

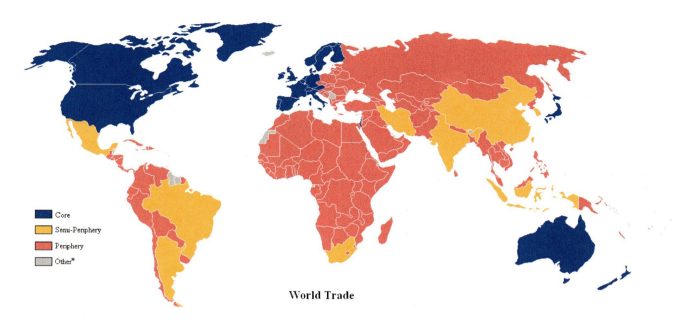

Figure 1.6: The Global and Periphery (Map by Lou Coban, Wikimedia Commons, Public Domain)

1.4 THE PHYSICAL SETTING

When we describe places, we can discuss their absolute and relative location and their relationship and interaction with other places. As regional geographers, we can dig deeper and explore both the physical and human characteristics that make a particular place unique. Geographers explore a wide variety of spatial phenomena, but the discipline can roughly be divided into two branches: physical geography and human geography. Physical geography focuses on natural features and processes, such as landforms, climate, and water features. Human geography is concerned with human activity, such as culture, language, and religion. However, these branches are not exclusive. You might be a physical geographer who studies hurricanes, but your research includes the human impact from these events. You might be a human geographer who studies food, but your investigations include the ecological impact of agricultural systems. Regional geography takes this holistic approach, exploring both the physical and human characteristics of the world's regions.

Much of Earth's physical landscape, from mountains to volcanoes to earthquakes to valleys, has resulted from the movement of tectonic plates. As the theory of **plate tectonics** describes, these rigid plates are situated on top of a bed of molten, flowing material, much like a cork floating in a pot of boiling water. There are seven major tectonic plates and numerous minor plates (see **Figure 1.7**).

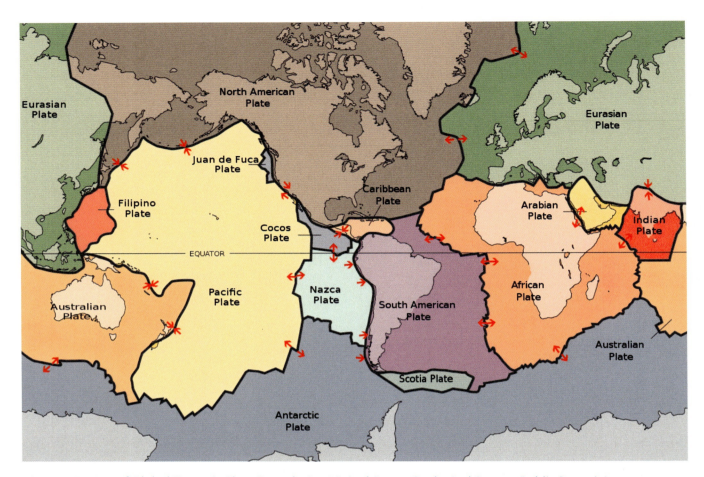

Figure 1.7: Map of Global Tectonic Plate Boundaries (United States Geological Survey, Public Domain)

Where two tectonic plates meet is known as a plate boundary and boundaries can interact in three different ways (see **Figure 1.8**). Where two plates slide past one another is called a transform boundary. The San Andreas Fault in California is an example of a transform boundary. A divergent plate boundary is where two plates slide apart from one another. Africa's Rift Valley was formed by this type of plate movement. Convergent plate boundaries occur when two plates slide towards one another. In this case, where two plates have roughly the same density, upward movement can occur, creating mountains. The Himalaya Mountains, for example, were formed from the Indian plate converging with the Eurasian plate. In other cases, subduction occurs and one plate slides below the other. Here, deep, under-ocean trenches can form. The 2004 Indian Ocean earthquake and tsunami occurred because of a subducting plate boundary off the west coast of Sumatra, Indonesia.

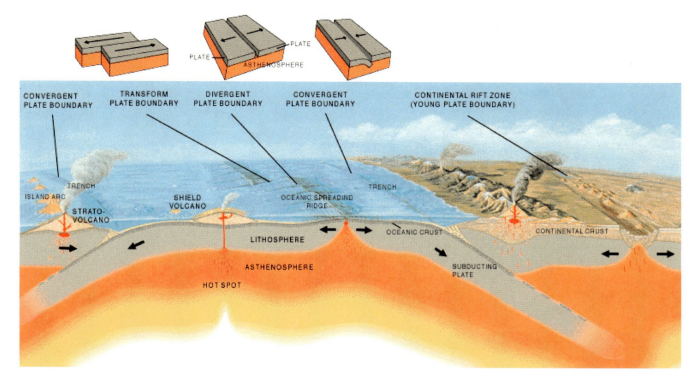

Figure 1.8: Types of Tectonic Plate Boundaries (United States Geological Survey, Public Domain)

Interaction between tectonic plates and historical patterns of erosion and deposition have generated a variety of landforms across Earth's surface. Each of the world's regions has identifiable physical features, such as plains, valleys, mountains, and major water bodies. Topography refers to the study of the shape and features of the surface of the Earth. Areas of high relief have significant changes in elevation on the landscape, such as steep mountains, while areas of low relief are relatively flat.

Another key feature of Earth's physical landscape is climate. Weather refers to the short-term state of the atmosphere. We might refer to the weather as partly sunny or stormy, for example. Climate, on the other hand, refers to long-term weather patterns and is affected by a place's latitude, terrain, altitude, and nearby water bodies. Explained another way, "weather" is what you're wearing today while "climate" is all the clothes in your closet. Geographers commonly use the Köppen climate classification system to refer to the major climate zones found in the world (see **Figure 1.9**).

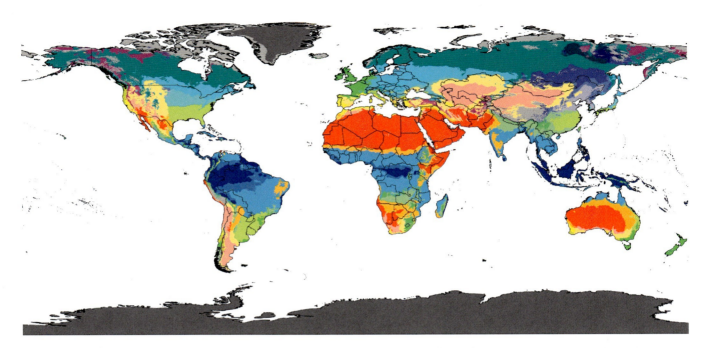

Figure 1.9: World Map of Köppen Climate Classifications (© Ali Zifan, Wikimedia Commons, CC BY 4.0)

Each climate zone in the Köppen climate classification system is assigned a letteredcode, referring to the temperature and precipitation patterns found in the particular region. Climate varies widely across Earth. Cherrapunji, India, located in the Cwb climate zone, receives over 11,000 mm (400 in) of rain each year. In contrast, the Atacama Desert (BWk), situated along the western coast of South America across Chile, Peru, Bolivia, and Argentina, typically receives only around 1 to 3 mm (0.04 to 0.12 in) of rain each year.

Earth's climate has gone through significant changes historically, alternating between long periods of warming and cooling. Since the industrial revolution in the 1800s, however, global climate has experienced a warming phase. 95 percent of scientists agree that this global **climate change** has resulted primarily from human activities, particularly the emission of greenhouse gases like carbon dioxide (see **Figure 1.10**). 17 of the 18 warmest years ever recorded have occurred since 2000. Overall, this warming has contributed to rising sea levels as the polar ice caps melt, changing precipitation patterns, and the expansion of deserts. The responses to global climate change, and the impacts from it, vary by region.

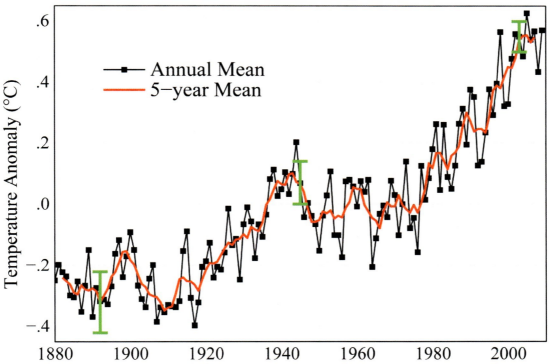

Figure 1.10: Mean Land-Ocean Surface Temperature Index, 1880 to Present (NASA, Public Domain)

1.5 THE HUMAN SETTING

The physical setting of the world's places has undoubtedly influenced the human setting, just as human activities have shaped the physical landscape. There are currently around 7.4 billion people in the world, but these billions of people are not uniformly distributed. When we consider where people live in the world, we tend to cluster in areas that are warm and are near water and avoid places that are cold and dry. As shown in **Figure 1.11**, there are three major population clusters in the world: East Asia, South Asia, and Europe.

Figure 1.11: Map of Global Population Clusters (Derivative work from original by Cocoliras, Wikimedia Commons)

Just as geographers can discuss "where" people are located, we can explore "why" population growth is occurring in particular areas. All of the 10 most populous cities in the world are located in countries traditionally categorized as "developing." These countries typically have high rates of population growth. A population grows, quite simply, when more people are born than die. The birth rate refers to the total number of live births per 1,000 people in a given year. In 2012, the average global birth rate was 19.15 births per 1,000 people.

Subtracting the death rate from the birth rate results in a country's **rate of natural increase** (RNI). For example, Madagascar has a birth rate of 32.9 per 1,000 (as of 2017) and a death rate of 6.19 per 1,000. 32.9 minus 6.19 is 26.71 per 1,000. If you divide the result by 10, you'd get 2.671 per 100, or 2.671 percent. In essence, this means that Madagascar's population is increasing at a rate of 2.671 percent per year. The natural increase rate does not include immigration. Some countries in Europe, in fact, have a negative natural increase rate, but their population continues to increase due to immigration.

The birth rate is directly affected by the **total fertility rate** (TFR), which is the average number of children born to a woman during her child-bearing years (see **Figure 1.12**). In developing countries, the total fertility rate is often 4 or more children, contributing to high population growths. In developed countries, on the other hand, the total fertility rate may be only 1 or 2 children, which can ultimately lead to population decline.

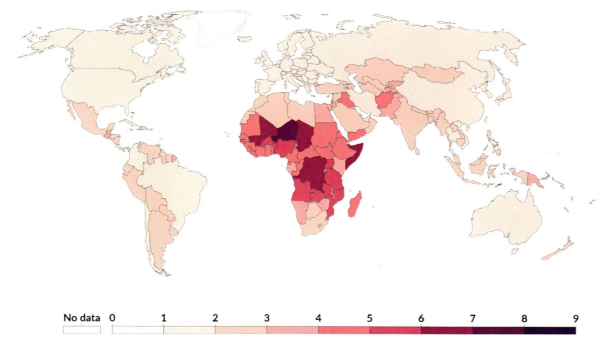

Figure 1.12: Map of Countries by Fertility Rate, 2015 (© Our World in Data, CC BY-SA)

A number of factors influence the total fertility rate, but it is generally connected to a country's overall level of development. As a country develops and industrializes, it generally becomes more urbanized. Children are no longer needed to assist with family farms, and urban areas might not have large enough homes for big families. Women increasingly enter the workforce, which can delay childbearing and further restrict the number of children a family desires. Culturally, a shift occurs as industrialized societies no longer value large family sizes. As women's education increases, women are able to take control of their reproductive rights. Contraceptive use becomes more widespread and socially acceptable.

This shift in population characteristics as a country industrialized can be represented by the **demographic transition model** (DTM) (see **Figure 1.13**). This model demonstrates the changes in birth rates, death rates, and population growth over time as a country develops. In stage one, during feudal Europe, for example, birth rates and death rates were very high. Populations were vulnerable to drought and disease and thus population growth was minimal. No country remains in stage one today. In stage two, a decline in death rates leads to a rise in population. This decline in death rates occurred as a result of agricultural productivity and improvements in public health. Vaccines, for example, greatly reduced the mortality from childhood diseases.

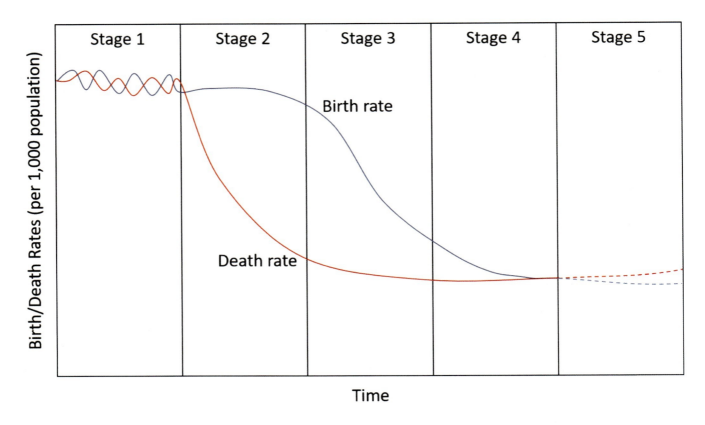

Figure 1.13: The Demographic Transition Model (Figure by author)

Stage two countries are primarily agricultural, and thus there is a cultural and historical preference for large families, so birth rates remain high. Most of Sub-Saharan Africa is in stage two. In stage three, urbanization and increasing access to contraceptives leads to a decline in the birth rate. As country industrializes, women enter the workforce and seek higher education. The population growth begins to slow. Much of Middle and South America as well as India are in stage three.

In stage four, birth rates approach the death rates. Women have increased independence as well as educational and work opportunities, and families may choose to have a small number of children or none at all. Most of Europe as well as China are in stage four. Some have proposed a stage five of the demographic transition model. In some countries, the birth rate has fallen below the death rate as families choose to only have 1 child. In these cases, a population will decline unless there is significant immigration. Japan, for example, is in stage five and has a total fertility rate of 1.41. Although this is only a model, and each country passes through the stages of demographic transition at different rates, the generalized model of demographic transition holds true for most countries of the world.

As countries industrialize and become more developed, they shift from primarily rural settlements to urban ones. **Urbanization** refers to the increased proportion of people living in urban areas. As people migrate out of rural, agricultural areas, the proportion of people living in cities increases. As people living in cities have children, this further increases urbanization. For most of human history, we have been predominantly rural. By the middle of 2009, however, the number of people living in urban areas surpassed the number of people living in rural areas for the first time. In 2014, 54 percent of the world's population lived in urban areas. This figure is expected to increase to 66 percent by 2050.

The number of megacities, cities with 10 million people or more, has also increased. In 1990, there were 10 megacities in the world. In 2014, there were 28 megacities. Tokyo-Yokohama is the largest metropolitan area in the world with over 38 million inhabitants.

1.6 THE WORLD'S REGIONS

The world can be divided into regions based on human and/or physical characteristics. Regions simply refer to spatial areas that share a common feature. There are three types of regions: formal, functional, and vernacular. **Formal regions**, sometimes called homogeneous regions, have at least one characteristic in common. A map of plant hardiness regions, as in **Figure 1.14**, for example, divides the United States into regions based on average extreme temperatures, showing which areas particular plants will grow well. This isn't to say that everywhere within a particular region will have the same temperature on a particular day, but rather that in general, a region experiences the same ranges of temperature. Other formal regions might include religious or political affiliation, agricultural crop zones, or ethnicity. Formal regions might also be established by governmental organizations, such as the case with state or provincial boundaries.

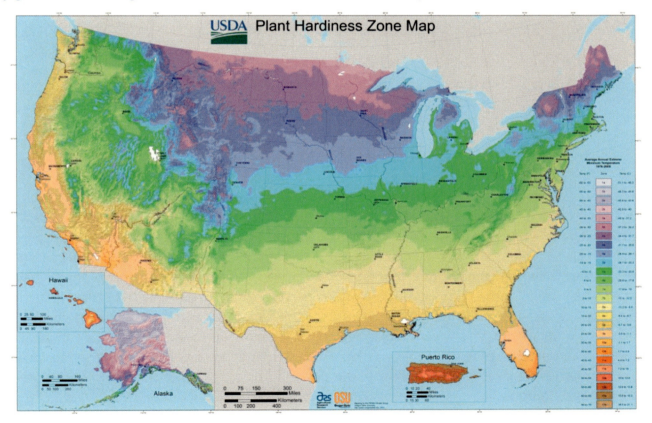

Figure 1.14: Map of Plant Hardiness Zones in the United States (USDA, Public Domain)

Functional regions, unlike formal regions, are not homogenous in the sense that they do not share a single cultural or physical characteristic. Rather, functional regions are united by a particular function, often economic. Functional regions are sometimes called nodal regions and have a nodal arrangement, with a core and surrounding nodes. A metropolitan area, for example, often includes a central city and its surrounding suburbs. We tend to think of the area as a "region" not because everyone is the same religion or ethnicity, or has the same political affiliation, but because it functions as a region. Los Angeles, for example, is the second-most populous city in the United States. However, the region of Los Angeles extends far beyond its official city limits as show in **Figure 1.15**. In fact, over 471,000 workers commute into Los Angeles County from the surrounding region every day. Los Angeles, as with all metropolitan areas, functions economically as a single region and is thus considered a functional region. Other examples of functional regions include church parishes, radio station listening areas, and newspaper subscription areas.

Figure 1.15: Map of Los Angeles Metro Area (© Kmusser, Wikimedia Commons, CC BY-SA 3.0)

Vernacular regions are not as well-defined as formal or functional regions and are based on people's perceptions. The southeastern region of the United States is often referred to as "the South," but where the exact boundary of this region is depends on individual perception (see **Figure 1.16**). Some people might include all of the states that formed the Confederacy during the Civil War. Others might exclude Missouri or Oklahoma. Vernacular regions exist at a variety of scales. In your hometown, there might be a vernacular region called "the west side." Internationally, regions like the Midlands in Britain or the Swiss Alps are considered vernacular. Similarly "the Middle East" is a vernacular region. It is perceived to exist as a result of religious and ethnic characteristics, but people wouldn't necessarily agree on which countries to include. Vernacular regions are *real* in the sense that our perceptions are real, but their boundaries are not uniformly agreed upon.

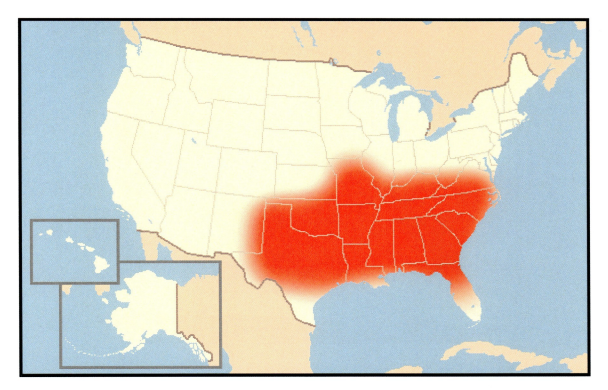

Figure 1.16: Map of the US "South" Vernacular Region (© Qz10, Wikimedia Commons, CC BY-SA 3.0)

As geographers, we can divide the world into a number of different regions based upon formal criteria and functional interaction. However, there is a matter of perception, as well. We might divide the world based on landmasses, since landmasses often share physical and cultural characteristics. Sometimes water connects people more than land, though. In the case of Europe, for example, the Mediterranean Sea historically provided economic and cultural links to the surrounding countries though we consider them to be three separate continents. Creating regions can often be a question of "lumpers and splitters;" who do you lump together and who do you split apart? Do you have fewer regions united by only a couple characteristics, or more regions that share a great deal in common?

This text takes a balanced approach to "lumping and splitting," identifying nine distinct world regions (see **Figure 1.17**). These regions are largely vernacular, however. Where does "Middle" America end and "South" America begin, and why is it combined into a single region? Why is Pakistan, a predominantly Muslim country, characterized as "South" Asia and not "Southwest" Asia? Why is Russia its own region? You might divide the world into entirely different regions, maybe seven based on the continents, or just two: the "core" and the "periphery." These nine regions are not universally agreed-upon; they are simply foundations for discussing the different areas of the world.

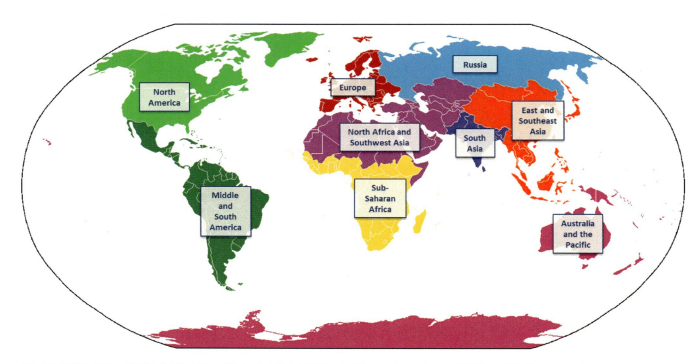

Figure 1.17: Map of World Regions (Image adapted from Cogito ergo sumo, Wikimedia Commons)

Furthermore, while it might seem like there are clear boundaries between the world's regions, in actuality, where two regions meet are zones of gradual transition. These **transition zones** are marked by gradual spatial change. Moscow, Russia, for example, is quite similar to other areas of Eastern Europe, though they are considered two different regions on the map. Were it not for the Rio Grande and a large border fence dividing the cities of El Paso, Texas and Ciudad Juárez, Mexico, you might not realize that this metropolitan area stretches across two countries and world regions. Even within regions, country borders often mark spaces of gradual transition rather than a stark delineation between two completely different spaces. The border between Peru and Ecuador, for example, is quite relaxed as international borders go and residents of the countries can move freely across the boundary to the towns on either side (see **Figure 1.18**).

Figure 1.18: Sign Welcoming People Entering Peru from Ecuador (© Vanished_user_j123kmqwfk56jd, Wikimedia Commons, CC BY-SA 3.0)

1.7 SUB-DISCIPLINES OF GEOGRAPHY

Geography has two primary branches, physical and human geography, but numerous sub-disciplines, many of which include both physical and human elements. Furthermore, as with world regions, it's often difficult to make precise boundaries between fields of study. A geographer might be a human geographer who specializes in culture who further specializes in religion. That same geographer might also conduct side research on environmental issues. And she might, in her spare time, investigate geographies of fictional landscapes. One benefit of geography is that its breadth offers a wide array of phenomena to explore. Everything happens somewhere, and thus everything is geographical.

Within physical geography, the main sub-disciplines are: biogeography (the study of the spatial distribution of plants and animals), climatology (the study of climate), hydrology (the study of water), and geomorphology (the study of Earth's topographic features). This list is not inclusive, however. Some geographers study geodesy, the scientific measurement and representation of

Earth. Others study pedology, the exploration of soils. What unites physical geographers is an emphasis on the scientific study of the physical features of Earth in all of its many forms.

Human geography, too, consists of a number of sub-disciplines that often overlap and interact. The main sub-disciplines of human geography include: cultural geography (the study of the spatial dimension of culture), economic geography (the study of the distribution and spatial organization of economic systems), medical geography (the study of the spatial distribution of health and medicine), political geography (the study of the spatial dimension of political processes), population geography (also known as demography, the study of the characteristics of human populations), and urban geography (the study of urban systems and landscapes). Human geographers essentially explore how humans interact with and affect the earth.

Political geography provides the foundation for investigating what many people understand as geography: countries and governmental structures. Political geographers ask questions like "Why does a particular state have a conflict with its neighbor?" and "How does the government of a country affect its voting patterns?" When political geographers study the world, they refer to **states**, which are independent, or sovereign, political entities recognized by the international community. States are commonly called "countries" in the United States; Germany, France, China, and South Africa are all "states." So how many states are there in the world? The question is not as easy to answer as it might seem. What if a state declares itself independent, but is not recognized by the entirety of the international community? What if a state collaborates so closely with its neighbor that it gives up some of its sovereignty? What happens if a state is taken over by another state? As of 2019, there are 206 states that could be considered sovereign, though some are disputed and are only recognized by one other country. Only 193 states are members of the United Nations. Others, like Palestine, are characterized as "observer states." The United States Department of State recognizes 195 states as independent, including the Holy See, often known as Vatican City, and Kosovo, a disputed state in Southeastern Europe.

In addition to questions of sovereignty, political geographers investigate the various forms of government found around the world (see **Figure 1.19**). States govern themselves in a variety of ways, but the two main types of government are unitary and federal. In a **unitary state**, the central government has the most power. Local or regional governments might have some decision-making power, but only at the command of the central government. Most of the states of the world have unitary systems. A **federal state**, on the other hand, has numerous regional governments or self-governing states in addition to a national government. Several large states like the United States, Russia, and Brazil are federations.

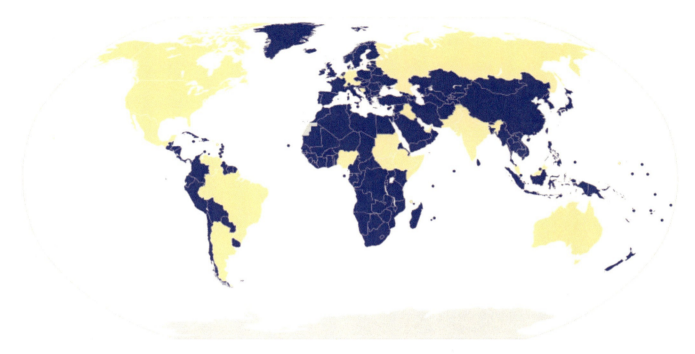

Figure 1.19: Map of Federal and Unitary States (Derivative work from original by Lokal_Profil, Wikimedia Commons)

Economic geographers explore the spatial distribution of economic activities. Why are certain states wealthier than others? Why are there regional differences related to economic development within a country? All countries have some sort of economic system but have different resources, styles of development, and government regulations. So how can we compare countries in terms of economic development? One way is by examining a country's **gross domestic product** (GDP), the value of all the goods and services produced in a country in a given year. Often, it is helpful to divide GDP by the number of people in a country; this is known as GDP per capita and it roughly equates to average income. However, goods have different costs in different countries, so GDP per capita is generally given in terms of purchasing power parity (PPP), meaning that each country's currency is adjusted so that it has roughly the same purchasing power. Thus, GDP per capita in terms of PPP simply refers to the amount of goods and services produced in a country divided by the number of people in that country and then adjusted for how much goods and services actually cost in that country.

One limitation of GDP is that it only takes into account the goods and services produced domestically. However, many businesses today have locations and production facilities in other countries. **Gross national income** (GNI) is a way to measure a country's economic activity that includes all the goods and services produced in a country (GDP) as well as income received from overseas.

1.8 GLOBALIZATION AND INEQUALITY

When we start to explore the spatial distribution of economic development, we find that there are stark differences between and within world regions. Some countries have a very high standard of living and high average incomes, while others have few resources and high levels of poverty. Politically, some countries have stable, open governments, while others have long-standing authoritarian regimes. Thus, world regional geography is, in many ways, a study of global inequality. But the geographic study of inequality is more than just asking where inequalities are present, it is also digging deeper and asking why those inequalities exist.

How can we measure inequality? Generally, inequality refers to uneven distributions of wealth, which can actually be challenging to measure. By some accounts, the wealthiest one percent of people in the world have as much wealth as the bottom 99 percent. Wealth inequality is just one facet of global studies of inequality, however. There are also differences in income: around half of the world survives on less than $2 per day, and around one-fifth have less than $1 per day (see **Figure 1.20**). There are also global differences in literacy, life expectancy, and healthcare. There are differences in the rights and economic opportunities for women compared to men. There are differences in the way resources are distributed and conserved.

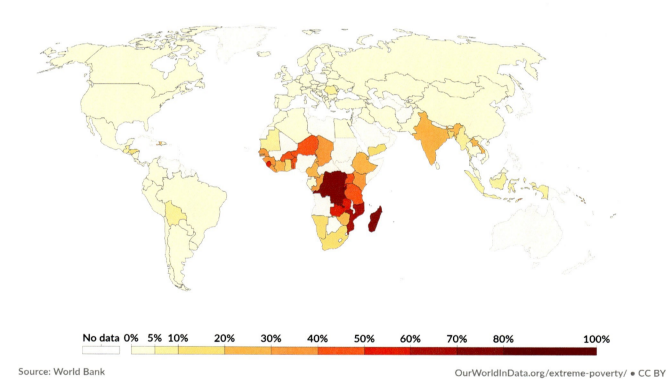

Figure 1.20: Share of the Population Living in Extreme Poverty (World Bank Estimates, 2017) (© Our World in Data, CC BY-SA)

Furthermore, these differences don't exist in a bubble. The world is increasingly interconnected, a process known as **globalization**. This increased global integration is economic but also cultural. An economic downturn in one country can affect its trading partners half a world away. A Hollywood movie might be translated in dozens of different languages and distributed worldwide.

Today, it is quite easy for a businesswoman in the United States to video chat with her factory manager in a less developed country. For many, the relative size of the world is shrinking as a result of advances in transportation and communications technology.

For others, though, particularly those in the poorest, most debt-ridden countries, the world is not flat. As global poverty rates have decreased over the past few decades, the number of people living in poverty within Sub-Saharan Africa has increased. In addition, while global economic integration has increased, most monetary transactions still occur within rather than between countries. The core countries can take advantage of globalization, choosing from a variety of trading partners and suppliers of raw materials, but the same cannot always be said of those in the periphery. Globalization has often led to cultural homogenization, as "Western" culture has increasingly become the global culture. American fast food chains can now be found in a majority of the world's countries. British and American pop music plays on radio stations around the world. The Internet in particular has facilitated the rapid diffusion of cultural ideas and values. But how does globalization affect local cultures? Some worry that as global culture has become more homogenized, local differences are slowly erasing. Traditional music, clothing, and food preferences might be replaced by foreign cultural features, which can lead to conflict. There is thus a tension between globalization, and the benefits of global connectivity, and local culture.

It is the uniqueness of the world's regions, the particular combination of physical landscapes and human activities, that has captivated geographers from the earliest explorers to today's researchers. And while it might simply be interesting to read about distant cultures and appreciate their uniqueness, geographers continue to dig deeper and ask why these differences exist. Geography matters. Even as we have become more culturally homogeneous and economically interconnected, there remain global differences in the geography of countries and these differences can have profound effects. Geographic study helps us understand the relationship between the world's communities, explain global differences and inequalities, and better address future challenges.

CHAPTER 2

Europe

> **Learning Objectives**
>
> - Identify the key geographic features of Europe
> - Explain how the industrial revolution has shaped the geographic landscape of Europe
> - Summarize how migration has impacted Europe's population
> - Describe the current controversies regarding migration to Europe

2.1 EUROPEAN PHYSICAL GEOGRAPHY AND BOUNDARIES

Europe? Where's that? It might seem like a relatively easy question to answer, but looking at the map, the boundaries of Europe are harder to define than it might seem. Traditionally, the continent of "Europe" referred to the western extremity of the landmass known as Eurasia (see **Figure 2.1**). Eurasia is a massive tectonic plate, so determining where exactly Europe ends and Asia begins is difficult. Europe is bordered by the Arctic Ocean in the North, the Atlantic Ocean and its seas to the west, and the Mediterranean Sea and the Black Sea to the south. Europe's eastern boundary is typically given as the Ural Mountains, which run north to south from the Arctic Ocean down through Russia to Kazakhstan. The western portion of Russia, containing the cities of St. Petersburg and Moscow, is thus considered part of Europe while the eastern portion is considered part of Asia. Culturally and physiographically, Western Russia is strikingly similar to Eastern Europe. These two regions share a common history as well with Russian influence extending throughout this transition zone.

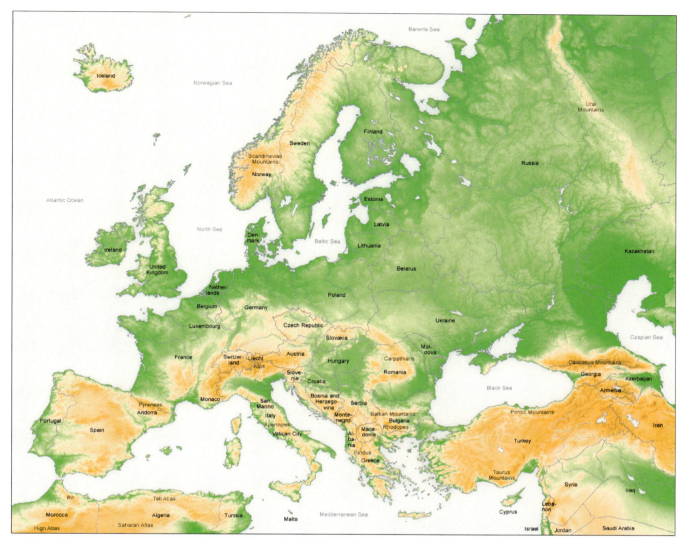

Figure 2.1: European Physical Geography and Political Boundaries (© San Jose, Wikimedia Commons, CC BY-SA 3.0)

In addition to the Ural Mountains, Europe has several other mountain ranges, most of which are in the southern portion of the continent. The Pyrenees, the Alps, and the Carpathians divide Europe's southern Alpine region from the hilly central uplands. Northern Europe is characterized by lowlands and is relatively flat. Europe's western highlands include the Scandinavian Mountains of Norway and Sweden as well as the Scottish Highlands.

Europe has a large number of navigable waterways, and most places in Europe are relatively short distances from the sea. This has contributed to numerous historical trading links across the region and allowed for Europe to dominate maritime travel. The Danube, sometimes referred to as the "Blue Danube" after a famous Austrian waltz of the same name, is the European region's largest river and winds its way along 2,860 km (1,780 mi) and 10 countries from Germany to Ukraine.

This proximity to water also affects Europe's climate (see **Figure 2.2**). While you might imagine much of Europe to be quite cold given its high latitudinal position, the region is surprisingly temperate. The Gulf Stream brings warm waters of the Atlantic Ocean to Europe's coastal region and warms the winds that blow across the continent. Amsterdam, for example, lies just above the 52°N line of latitude, around the same latitudinal position as Saskatoon, in Canada's central Saskatchewan province. Yet Amsterdam's average low in January, its coldest month, is around 0.8°C (33.4°F) while Saskatoon's average low in January is -20.7°C (-5.3°F)!

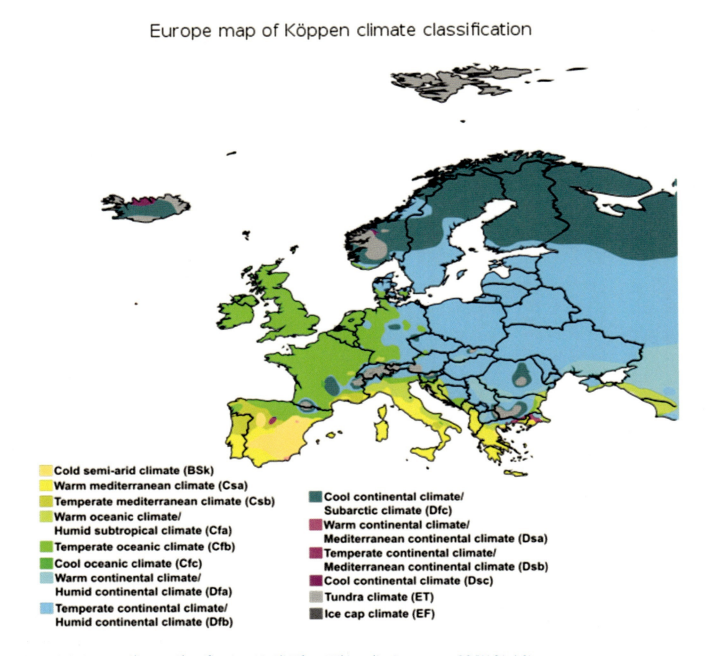

Figure 2.2: Europe Climate Classification (© Ali Zifan, Wikimedia Commons, CC BY-SA 4.0)

While geographers can discuss Europe's absolute location and the specific features of its physical environment, we can also consider Europe's relative location. That is, its location relative to other parts of the world. Europe lies at the heart of what's known as the land hemisphere. If you tipped a globe on its side and split it so that half of the world had most of the land and half had most of the water, Europe would be at the center of this land hemisphere (see **Figure 2.3**). This, combined with the presence of numerous navigable waterways, allowed for maximum contact between Europe and the rest of the world. Furthermore, distances between countries in Europe are relatively small. Paris, France, for example, is just over a two-hour high speed rail trip from London, England.

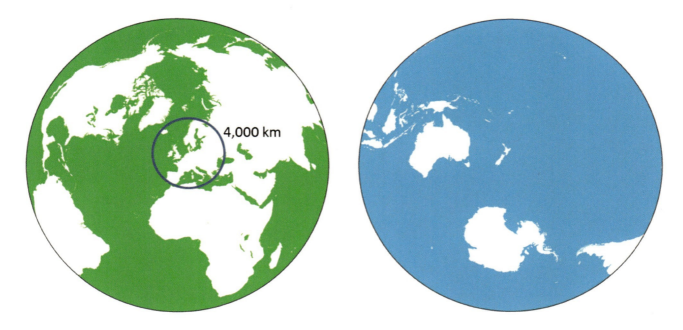

Figure 2.3: Map of Land and Water Hemispheres and Europe's Relative Location (Derivative work from original by Citynoise, Wikimedia Commons, CC BY-SA 4.0)

This relative location provided efficient travel times between Europe and the rest of the world, which contributed to Europe's historical dominance. When we consider globalization, the scale of the world is shrinking as the world's people are becoming more interconnected. For Europe, however, the region's peoples have long been interconnected with overlapping histories, physical features, and resources.

2.2 COOPERATION AND CONTROL IN EUROPE

Europe's physical landforms, climate, and underlying resources have shaped the distribution of people across the region. When early humans began settling this region, they likely migrated through the Caucasus Mountains of Southwest Asia and across the Bosporus Strait from what is now Turkey into Greece. The Greeks provided much of the cultural and political foundations for modern European society. Greek ideals of democracy, humanism, and rationalism reemerged in Europe during the Age of Enlightenment. The Roman Empire followed the Greek Empire, pushing further into Europe and leaving its own marks on European society (see **Figure 2.4**). Modern European architecture, governance, and even language can be traced back to the Roman Empire's influence.

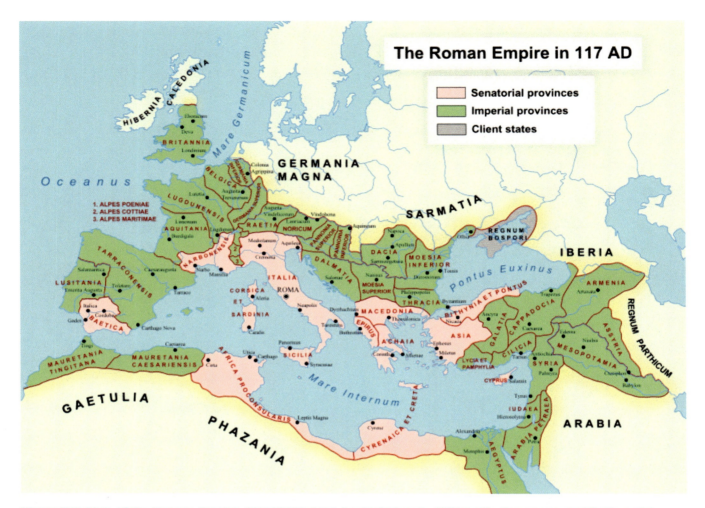

Figure 2.4: Map of the Roman Empire, 117 AD/CE (Map by Andrei nacu, Wikimedia Commons, Public Domain)

The Roman's vast European and Southwest Asian empire united the region under Christianity and created new networks of roads and trading ports. With the fall of the Roman Empire, however, tribal and ethnic allegiances reemerged and a number of invasions and migrations occurred. England, for example, was settled by the Germanic Anglo-Saxons, from which the name "England" or "Angeln" is derived, then by the Normans from present-day France.

Europe today is comprised of 40 countries, but historically, this was a region dominated by kingdoms and empires – even fairly recently. A map of Europe from just 200 years ago looks strikingly different from today's political boundaries (see **Figure 2.5**) . At that time, Greece and Turkey were still controlled by the Ottoman Empire and Italy was a conglomerate of various city-states and independent kingdoms. Many of the countries and political boundaries of Europe we know today were not formed until after World War II.

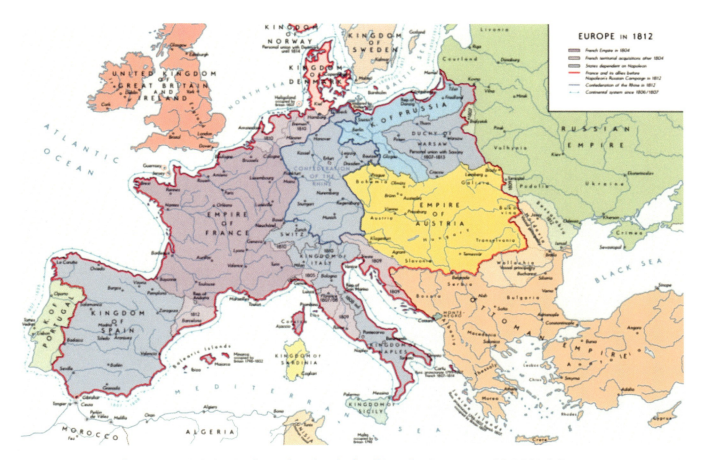

Figure 2.5: Map of Europe, 1812 CE (© Alexander Altenhof, Wikimedia Commons, CC BY-SA 3.0)

Europe's population has shifted and changed over time as well. Whereas Europe was once largely feudal and agrarian, today around 75 percent of its people live in cities. Europe's largest city is London, with a population of around 8.5 million within its city limits. Although the United Kingdom was the dominant force in Europe during industrialization, Germany now dominates the region in terms of population, gross domestic product, and size.

The political map of Europe continues to change, with shifting alliances, competing goals, and new pushes for independence. In general, Western Europe has moved toward cooperation. The European Union developed out of the Benelux Economic Union signed in 1944 between Belgium, the Netherlands, and Luxembourg. France, Italy, and West Germany signed an economic agreement with the Benelux states in 1957, and from there, the economic cooperation continued to expand. The European Union (EU) itself was created in 1993 and today, the organization has 27 member countries (see **Figure 2.6**). Not all members of the EU use the euro, its official currency; the 19 member states who do are known as the eurozone. The United Kingdom is the only state to have left the EU. Its withdrawal is known as "Brexit" (meaning "British exit") and was finalized in January 2020.

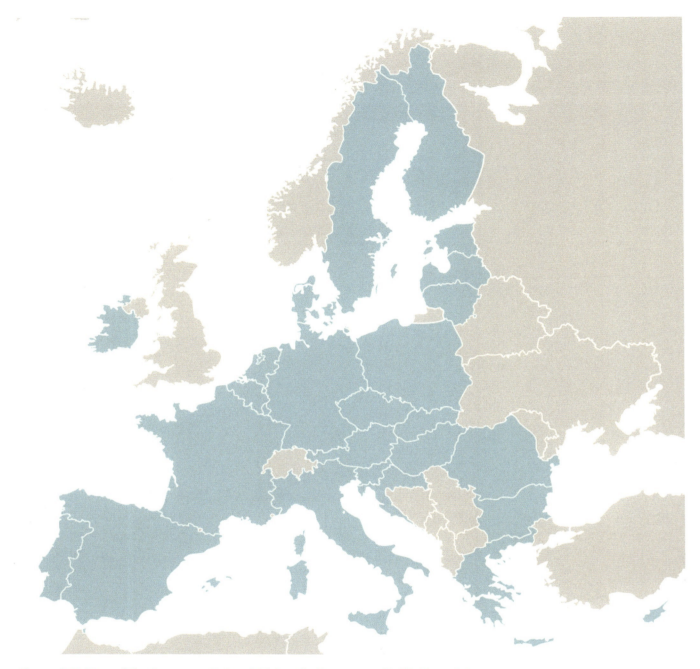

Figure 2.6: Map of the European Union (Wikimedia Commons, Public Domain)

Today, it is relatively easy to travel across Europe, in part because of economic and monetary cooperation, but also because internal border checks have largely been abolished. The Schengen Agreement, signed in the 1990s allows member states to essentially function as a single territory in terms of entry. These states share a common visa system and residents and vehicles can travel freely throughout states participating in the agreement.

Although the European Union has provided member states with a number of advantages, the system has had some structural concerns. Greece, for example, admitted to the EU in 1981, adopted the euro in 2001. It has had continued issues with debt, however, and has required massive bailouts from other member states. The United Kingdom held a referendum in June 2016 and decided to leave the EU, the first time a country has made the decision to leave the organization. When analyzing the EU and its advantages and disadvantages, you might consider why a country would join a supranational organization. To join an organization like the EU, a country gives up some of its sovereignty, its independence in making economic, political, or legal decisions. Ideally, a country would gain more than it loses. Countries united economically can more easily facilitate trade, for example, or could share a common military rather than each supporting their own. Those who favored the United Kingdom withdrawing from the EU, however, argued that membership in the EU did not offer enough advantages and preferred the United Kingdom to control its own trade deals and immigration restrictions.

Devolution, which occurs when regions within a state seek greater autonomy, has continued in Europe, representing a tension between nationalistic ideals and ethnic ties. In the United Kingdom, a 2014 Scottish independence referendum was narrowly defeated but led to greater autonomy for Scotland. In general, policies offering increased autonomy have kept the map of Western Europe fairly intact. Ethnic groups seeking sovereignty often want political autonomy but economic integration, and thus devolution generally allows them more decision-making power.

In the Balkan region, however, strong ethnic identities has contributed to continued political instability and the formation of new states. In fact, the devolutionary forces found in this region led to the creation of the term Balkanization, referring to the tendency of territories to break up into smaller, often hostile units. The Balkans came under the control of the Ottoman Empire, and once the empire collapsed following World War I, several territories in this region were joined together as the country of Yugoslavia (see **Figure 2.7**). Following World War II, Yugoslavia was led by Josip Broz Tito who attempted to unify the region by suppressing ethnic allegiances in favor of national unity. After his death, however, those ethnic tensions reemerged. In the 1990s, Yugoslavia was led by the dictator Slobodan Milošević, a Serbian who supported a genocidal campaign against the region's Croats, Bosnians, and Albanians. In Bosnia alone, over 8,000 Bosnian Muslim men and boys were killed. Most recently, Kosovo, comprised mostly of Albanian Muslims, declared independence from Serbia in 2008, though its status as a sovereign state is still contested by some, including Serbia, Bosnia, and Greece.

Figure 2.7: Map of Former Yugoslavia (United Nations, Public Domain)

The map of Europe continues to evolve. In February 2019, for instance, the country formerly known as the Former Yugoslav Republic of Macedonia officially changed its name to the Republic of North Macedonia, or just North Macedonia, resolving a long dispute with Greece.

While some countries in the region have decidedly benefitted from globalization, others remain fairly limited in terms of global trade and global economic integration. **Figure 2.8**, a map of gross domestic product (GDP) per capita reveals a marked difference between the states of Western Europe and the eastern region. Germany's GDP per capita as of 2017, for example, was $44,470 (in US dollars), according to the World Bank. In Moldova, a former Soviet republic bordering Romania, that figure was $2,290.

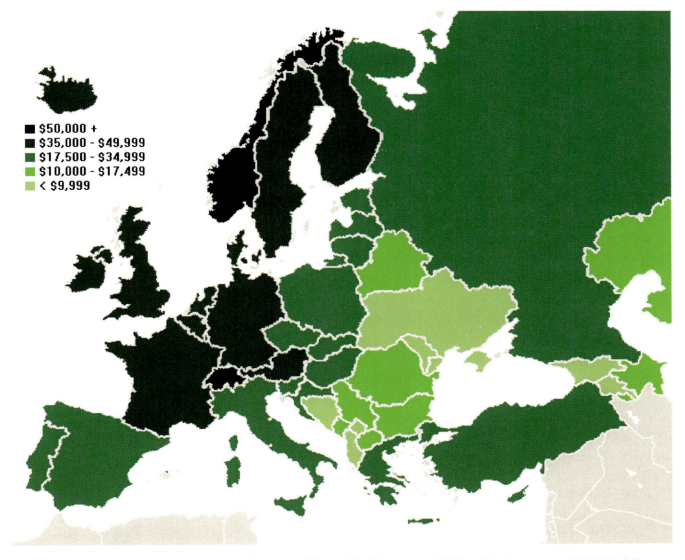

Figure 2.8: GDP (in PPP) per capita, 2012 from WorldBank (© XNeverEver, Wikimedia Commons, CC BY-SA 3.0)

2.3 THE INDUSTRIAL REVOLUTION

The differences in levels of development across Europe today have largely been shaped by the Industrial Revolution. The **Industrial Revolution** refers to the changes in manufacturing that occurred in the late 18th and early 19th centuries. These changes had profound effects on society, economics, and agriculture, not just in Europe, but globally.

Prior to the Industrial Revolution, most goods in Europe were produced in the home. These so-called "cottage" industries consisted of individual workers making unique goods in their homes, usually on a part-time basis. These products, such as clothing, candles, or small housewares, could be sold by a farming family to supplement their income.

The Industrial Revolution began in the United Kingdom, and while it's difficult to pinpoint the exact point at which the revolution began, a key invention was James Watt's steam engine, which entered production in 1775. This steam-driven engine was adopted by industries to allow for factory production. Machines could now be used instead of human or animal labor. Interestingly, a side effect of the steam engine was that it enabled better iron production, since iron required an even and steady stream of heat. This improved iron was then used to build more efficient steam engines, which in turn produced increasingly better iron. These improvements and new technologies gradually spread across Europe, eventually diffusing to the United States and Japan.

During this time, there were also significant changes in agricultural production. The Agrarian Revolution began in the mid-1750s and was based upon a number of agricultural innovations. This was the Age of Enlightenment, and the scientific reasoning championed during this era was applied to the growing of crops. Farmers began using mechanized equipment, rather than relying solely on human or animal labor. Fertilizers improved soil conditions, and crop rotation and complementary planting further increased crop yields. During this time period, there was a shift to commercial agriculture, where excess crops are sold for a profit, rather than subsistence agriculture, where farmers primarily grow food for their own family's consumption.

Around the same time improvements in rail transportation changed both the way goods were distributed across Europe and the movement of people across the region. The use of steam engines and improved iron also transformed the shipping industry, with steamships beginning to set sail across the Atlantic Ocean.

The Agrarian Revolution coupled with the Industrial Revolution profoundly changed European geography. With the improvements of the Agrarian Revolution, farmers could produce more with less work. This provided an agricultural surplus, enabling a sustained population increase. Port cities and capital cities became centers of trade and expanded. Critically, the Agrarian Revolution freed workers from having to farm, since fewer farmers were needed to produce the same amount of crops, enabling people to find work in the factories. These factories were primarily located in cities, and thus it was the combination of these two revolutions that dramatically increased urbanization in Europe. At the start of the 19th century, around 17 percent of England's population lived in cities; by the end of the 19th century, that figure had risen to 54 percent.

Overall, the Industrial Revolution considerably improved European power by boosting their economies, improving their military technology, and increasing their transportation efficiency. Even before the Industrial Revolution, Europe exerted a considerable amount of control over the rest of the world. European colonialism began in the 1400s, led by Portugal and Spain. In the 1500s, England, France, and the Dutch began their own colonial campaigns. By the start of WorldWar I, the British Empire, boosted by the improvements of the Industrial Revolution, had the largest empire in history, covering 20 percent of the world's population at the time (see **Figure 2.9**).

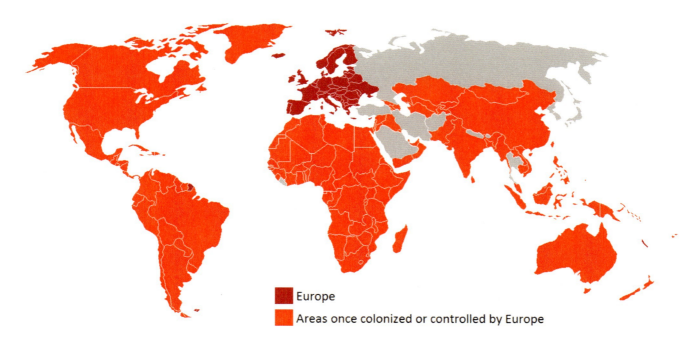

Figure 2.9: European Colonization (Derivative work from original by Cogito ergo sumo, Wikimedia Commons)

Coinciding with the Industrial and Agrarian Revolutions were a number of political revolutions in Europe. The most influential political change came as the result of the French Revolution, which occurred between 1789 and 1799 CE. This revolution ended France's monarchy, establishing a republic, and provided the foundation for numerous political revolutions that followed. It also weakened the power of the Roman Catholic Church in France, inspiring the modern-day separation between church and state that is typical of many Western countries, including the United States.

Today, the map of Europe reflects the changes brought about by the Industrial and Agrarian Revolutions as well as the political changes that took place throughout the time period. Europe's core area, where economic output is highest, is largely centered around the manufacturing areas that arose during the Industrial Revolution (see **Figure 2.10**). These manufacturing areas in turn were originally located near the raw materials, such as coal, that could sustain industrial growth.

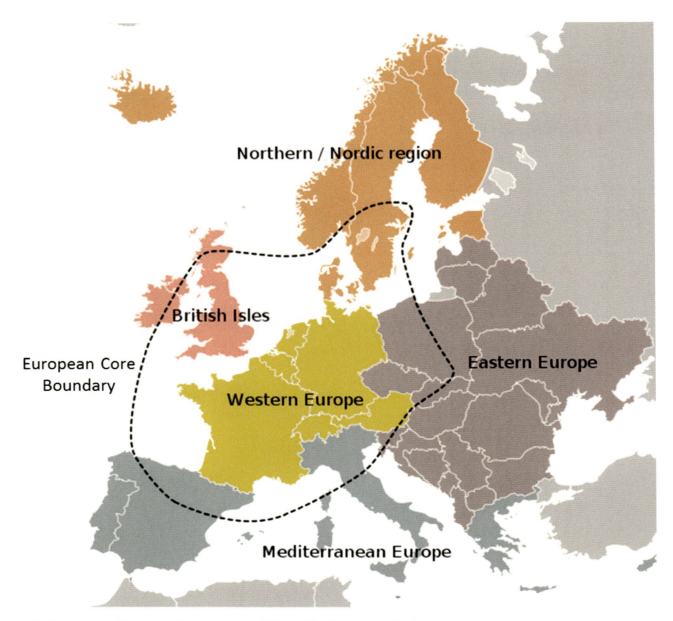

Figure 2.10: Regions of Europe (Map by Koyos, Wikimedia Commons, Public Domain)

The shift in labor that occurred during the Industrial Revolution, as people left rural farms to find work in factories, led to the specialization of labor that is found in Europe today. Areas within Europe tended to specialize in the production of particular goods. Northern Italy for example, has maintained a specialty in the production of textiles. Germany continues to specialize in automotive manufacturing. The Benz Patent-Motorwagen, the world's first gasoline powered automobile, was first built in Germany in 1886 and would later develop into the Mercedes-Benz corporation. As regions focused on the manufacture of particular goods, they benefited from **economies of scale**, the savings in cost per unit that results from increasing production. If you wanted to build a chair, for example, you'd have to buy the wood, glue, and screws as well as the tools needed to construct it, such as a drill, sander, and saw. That single chair would be quite costly to produce. If you wanted to make ten chairs, however, those same tools could be used for every chair, driving the cost of each chair down. Many areas in Europe have shifted from more traditional to high-tech manufacturing and industrial output in the region remains high.

2.4 EUROPEAN MIGRATION

The Industrial and Agricultural Revolutions shaped both migration patterns within Europe and immigration to the region. **Migration** refers to a move from one place to another intended to be permanent. When considering migration, geographers look at both intraregional migration, movement within a particular region, and interregional migration, such as migration from Europe to North America. Geographers who study migration also investigate push and pull factors that influence people to move. Push factors are those that compel you to move from your current location. A lack of job opportunities, environmental dangers, or political turmoil would all be considered push factors. Pull factors, on the other hand, are those that entice you to move to a new place, and might include ample jobs, freedom from political or religious persecution, or simply the availability of desirable amenities. Historically, most intraregional migration in Europe was rural to urban, as people moved from farms to cities to find work. Cities grew rapidly in the region as centers of trade and industry.

Before the industrial revolution, migration to the region was usually in the form of invasions, such as with the Roman Empire, the Islamic Empire, and the Ottoman Empire. One notable, historical migration that did not represent an invading empire was the Jewish diaspora following the conquest of Judea, the region now known as Israel and Palestine, by a number of groups including the Assyrians, Babylonians, and Romans. A diaspora refers to a group of people living outside of their ancestral homeland and many Jewish people moved to Europe to escape violence and persecution, particularly after the Roman destruction of the Second Temple in Jerusalem in 70 CE.

Jews migrating to Europe were often met with anti-Semitism, however. During the Middle Ages, European Jews were routinely attacked and were expelled from several countries including England and France. Jewish communities were destroyed during the mid-14th century as the Black Death swept across Europe and thousands of Jews were murdered, accused of poisoning the water and orchestrating the epidemic. In actuality, the disease was likely spread by rats, and worsened by the superstitious killing of cats in the same time period.

European Jews were often forced to live in distinct neighborhoods, also known as ghettos. In fact, this requirement to live in specific areas was required in Italy under areas ruled by the Pope until 1870. These distinctive communities were often met with suspicion by European Christians, many of whom continued to foster the same anti-Semitic sentiment that had been prevalent during the Middle Ages.

This anti-Semitic fervor and persecution of Jews reached its height at the time of the Nazi Party's rule in Germany. Prior to World War II, close to 9 million Jews lived in Europe; 6 million of them were killed in the Holocaust, the European genocide that targeted Jews, Poles, Soviets, communists, homosexuals, the disabled, and numerous other groups viewed as undesirable by the Nazi regime. Following the war, many surviving Jews emigrated back to the newly created state of Israel. Around 2.4 million Jews live in Europe today.

There was another shift in population after the signing of the Schengen Agreement in 1995, with large numbers of immigrants from Eastern Europe migrated to the western European countries in the core. Citizens of European Union countries are permitted to live and work in any country in the EU, and countries like the United Kingdom and Spain contain large numbers of Eastern European immigrants. Around half of all European migrants are from other countries within Europe.

Economic and political inequalities have driven much of the interregional migration to Europe since the 1980s. Immigrants from North Africa and Southwest Asia, for example, driven by limited employment opportunities and political conflict, have migrated to Europe in large numbers and now represent approximately 12 percent of all European migrants.

2.5 SHIFTING NATIONAL IDENTITIES

What does it mean to be European? Perhaps simply it means someone who's from Europe. But what does it mean to be French or German or Spanish or British? These countries have long been comprised of a number of different ethnic and linguistic groups (see **Figure 2.11**). Spain, for example, not only contains groups speaking Spanish, the language of the historic Castilian people of the region, but also the Basque-speaking region in the north, the Catalan-speaking region centered around Barcelona, and numerous other distinct language groups. The United Kingdom, while comprised primarily of people who identify as "English," also includes the areas of Wales, Scotland, and Northern Ireland, each with a distinct linguistic and cultural identity. The Welsh are actually believed to be the oldest ethnic group within the United Kingdom, so perhaps they could argue that they represent the original national identity.

Figure 2.11: Languages of Europe (Map by Andrei nacu, Wikimedia Commons, Public Domain)

Before the creation of states as we understand them today, Europe, as with the rest of the world, was divided largely by ethnicity or tribe. Empires often took control of multiple ethnic areas. Familial allegiances were of fundamental importance. That's not to say that geography or territory didn't matter, but simply that *who* you were mattered more than *where* you were.

The creation of sovereign political states changed this notion. Multiple ethnicities were often lumped together under single political entities, sometimes due to peaceful alliances and sometimes due to armed conquest. In cases where a state was dominated by a single, homogeneous ethnic and linguistic cultural identity, we would refer to it as a **nation-state**, from the term state, meaning a sovereign political area, and nation, meaning a group with a distinct ethnic and cultural identity. Several European countries today are considered nation-states, including Poland, where 93 percent of the population is ethnically Polish, and Iceland, which is 92 percent Icelandic. Historically, countries like France and Germany were also considered nation-states, though immigration has changed their cultural landscape.

The concept of nation-state is distinct from the idea of **nationalism**, which refers the feeling of political unity within a territory. National flags, anthems, symbols, and pledges all inspire a sense of belonging amongst people within a geographic area that is distinct from their ethnic identity.

What happens when feelings of nationalism and national identity are linked with a particular ethnic group? In cases where a particular ethnic group represents the majority, nationalist ideals might be representative of that group's language or religion. But what if there are other, minority ethnic groups that are excluded from what people think it means to be part of a particular state's nationalist identity?

Migration has continually changed the cultural landscape of Europe and as immigrant groups have challenged or been challenged by ideas of nationalism. In 1290 CE, King Edward I expelled all Jews from England, essentially establishing Christianity as being at the core of English national identity. This expulsion lasted until 1657 CE. In France after the French Revolution, ideas of nationalism included "liberty, equality, and fraternity," and extended into areas they conquered. In Germany, what it meant to be "German" under the Nazi Party excluded those who were considered to be "undesirable" and "enemies of the state," such as Jews, Roma (sometimes referred to as Gypsies), persons labeled as "homosexuals," communists, and others. Under Benito Mussolini, Italian nationalism excluded Slavs, Jews, and non-white groups.

Nationalism, taken to this extreme, is known as fascism. Fascists believe that national unity, to include a strong, authoritarian leader and a one-party state, provides a state with the most effective military and economy. Fascist governments might thus blame economic difficulty or military losses on groups that threaten national unity, even if those groups include their own citizens.

Within every country, ideas of nationalism grow, weaken, and change over time. **Centrifugal forces** are those that threaten national unity by dividing a state. These might include differing religious beliefs, linguistic differences, or even physical barriers within a state. **Centripetal forces**, on the other hand, tend to unify people within a country. A charismatic leader, a common religion or language, and a strong national infrastructure can all work as centripetal forces. Governments could also promote centripetal forces by unifying citizens against a common enemy, such as during the Cold War. Although the countries of Europe always had a significant amount of ethnic and linguistic variety, they typically maintained a strong sense of national identity. Religion in particular often worked as a centripetal force, uniting varying cultural groups under a common theological banner.

Religious adherence in Europe is shifting, however. In Sweden, for example, over 80 percent of the population belonged to the Church of Sweden, a Lutheran denomination, in 2000. By 2014, only 64.6 percent claimed membership in the church and just 18 percent of the population stated that they believed in a personal God (see Figure **2.12**). This is indicative of a broad shift in Europe from traditional, organized religion toward humanism or secularism. **Humanism** is a philosophy emphasizing the value of human beings and the use of reason in solving problems. Modern humanism was founded during the French Revolution, though early forms of humanism were integrated with religious beliefs. Secular humanism, a form of humanism that rejects religious beliefs, developed later. **Secularism** refers broadly to the exclusion of religious ideologies from government or public activities.

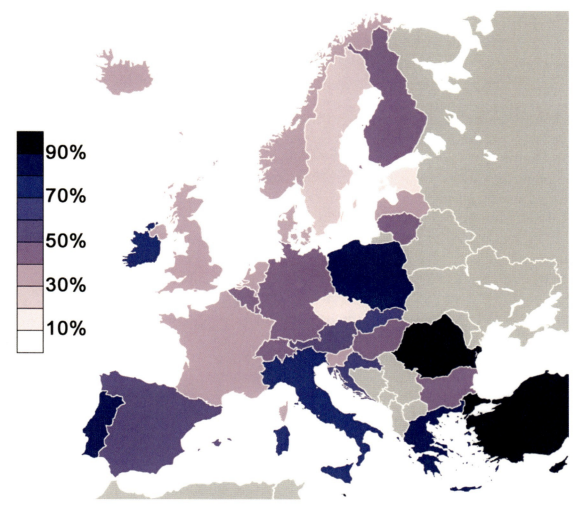

Figure 2.12: Percentage of People Who Answered "I believe there is a God" in 2005 Eurobarometer Poll (© Alphathon, Wikimedia Commons, CC BY-SA 3.0)

Geographers can examine how secularization has occurred in Europe; that is, how Europe has been transformed from countries with strong religious values to a more nonreligious society. In general, areas within the core of Europe tend to be more secular and thus some researchers link secularization with rising economic prosperity. Most Western European countries have strong social welfare programs, where citizens pay a higher percentage of taxes to support universal healthcare, higher education, child care, and retirement programs. These **social welfare** programs often serve as centripetal forces, unifying a country by providing government support and preventing citizens from falling into extreme poverty.

2.6 CURRENT MIGRATION PATTERNS AND DEBATES

The increasing secularization of Western Europe has magnified the conflict over immigration to the region. Whereas Western Europeans have become less religious over time, immigrants to the region are generally more religious. Increasing numbers of Muslim immigrants from North Africa and Southwest Asia have settled in Europe, lured by the hope of economic prosperity and political freedom. In 2010, around 6 percent of Europe identified as Muslim. That number is expected to grow to 10 percent by 2050. Muslims have the highest fertility rate among the major religious groups, so coupled with increasing immigration, this population is growing. In contrast, just under three-quarters of Europeans identified as Christian in 2010. This is expected to drop to 65 percent in 2050.

Europeans are divided about how open the region should be to immigrants, and how asylum seekers, refugees seeking sanctuary from oppression, should be treated. Even before the 2015 wave of Syrian migrants to Europe, a 2012-2014 survey showed that most Europeans (52 percent) wanted immigration levels to decrease. Opinions vary within the region, however. In the United Kingdom, 69 percent of people support decreased immigration. In Greece, a gateway country for migrants attempting to enter Europe, 84 percent of people desire decreased immigration. A majority of adults in Northern European countries, however, want immigration to stay the same or increase.

In 2014 and 2015, migration to Europe intensified as a result of an ongoing civil war in Syria. There were more refugees in 2014 than in any other year since World War II. 2015 shattered that record, however, as 65.3 million people were displaced. Germany has received the most applications for people seeking refugee status.

The journey for migrants is difficult and dangerous. Many attempt to cross by sea into Greece. Boats are often overcrowded and capsizing is common. Around 34 percent of refugees are children, many of them unaccompanied. Although the entire influx of refugees represents around 0.5 percent of Europe's population, it is not necessarily the sheer number of refugees that poses a problem, but rather, the *idea* of how immigrant populations might change the identity of a nation-state.

Many small towns in Europe have experienced shifting demographics as people move away to work in cities and immigrants move in to work in the available jobs. As Western Europe moved through industrialization, it has increasingly shifted away from heavy manufacturing and increased employment in service and high-tech industries, a process known as deindustrialization. The higher-skilled and higher-educated workers from small towns moved to the cities to find work, while lower-skilled immigrants worked the often dangerous or labor intensive jobs that remained. In the United Kingdom in particular, many of the people who oppose immigration and supported Britain leaving the EU are located in these small towns where immigration has quite visibly changed the cultural landscape that had already shifted as a result of deindustrialization.

For some, the debate over immigration and asylum are less questions of national identity and more issues of social justice. Do countries that have political freedom and economic prosperity have a moral obligation to assist those in need? Historically, the answer has often been "no." In 1938, on the brink of World War II, representatives from Western European countries voted not to accept Jewish refugees from Germany and Austria. Numerous countries in Europe have similarly voted not to accept Syrian migrants. Countries like Germany, which has accepted a relatively large number of asylum seekers, have been critical of other countries that have not been as welcoming. Sweden has specifically argued that if every country in Europe accepted a proportional amount of refugees, they would easily be able to accommodate the influx. Refugee populations typically have lower unemployment rates than native-born populations and though they require social services like housing and employment, can provide a long-term economic boost by increasing the labor force, especially in countries with otherwise declining populations.

Europe's population will continue to shift in terms of demographics and cultural identity. Recent economic changes and migration patterns have highlighted deep divides about ideas of national identity and the role of the region in global affairs. Europe continues to be an influential and economically important region and will likely continue to attract migrants from surrounding areas.

CHAPTER 3

Russia

> **Learning Objectives**
>
> - Identify the key geographic features of Russia
> - Analyze how the Russian Empire and the Soviet Union approached the issue of ethnic identity
> - Describe the current areas of ethnic conflict within Russia
> - Explain how Russian history impacted its modern-day geographic landscape

3.1 RUSSIA'S PHYSICAL GEOGRAPHY AND CLIMATE

Russia is the largest country in the world, containing 1/8 of the entire world's land area (see **Figure 3.1**). Russia is also the northernmost large and populous country in the world, with much of the country lying above the Arctic Circle. Its population, however, is comparatively small with around 143 million people, the majority of whom live south of the 60 degree latitude line and in the western portions of Russia near Moscow and Saint Petersburg. Russia stretches across eleven time zones, spanning 6,000 miles from Saint Petersburg on the Baltic Sea to Vladivostok on the Pacific Coast. The country also includes the exclave, or discontinuous piece of territory, of Kaliningrad situated between Poland and Lithuania.

Figure 3.1: Map of Russia (CIA World Factbook, Public Domain)

Because of its large size, Russia has a wide variety of natural features and resources. The country is located on the northeastern portion of the Eurasian landmass. It is bordered to the north by the Arctic Ocean, to the east by the Pacific Ocean, and to the south, by the Black and Caspian Seas. The Ural Mountains, running north to south, traditionally form the boundary between Europe and Asia and presented a formidable historical barrier to development. Culturally and physiographically, Western Russia, beyond the Ural Mountains, is quite similar to that of Eastern Europe. The region of Russia east of the Ural Mountains is known as Siberia.

In addition to the Ural Mountains, Russia contains several other areas of high relief (see **Figure 3.2**). Most notably, the Caucasus Mountains, forming the border between Russia and Southwest Asia, and the volcanic highlands of Russia's far east Kamchatka Peninsula. The western half of Russia is generally more mountainous than the eastern half, which is mostly low-elevation plains.

Figure 3.2: Topographical Map of Russia (© Tobias1984, Wikimedia Commons, CC BY-SA 3.0)

Russia's Volga River, running through central Russia into the Caspian Sea, is the longest river on the European continent and drains most of western Russia. The river is also an important source of irrigation and hydroelectric power. Lake Baikal, located in southern Siberia, is the world's deepest lake and also the world's largest freshwater lake. It contains around one-fifth of the entire world's unfrozen surface water. Like the deep lakes of Africa's rift valley, Lake Baikal was formed from a divergent tectonic plate boundary.

Although Russia's land area is quite large, much of the region is too cold for agriculture. As shown in **Figure 3.3**, the northernmost portion of Russia is dominated by **tundra**, a biome characterized by very cold temperatures and limited tree growth. Here, temperatures can drop below -50°C (-58°F) and much of the soil is **permafrost**, soil that is consistently below the freezing point of water (0°C or 32°F). South of the tundra is the **taiga** region, where coniferous, snow-capped forests dominate. This area of Russia contains the world's largest wood resources, though logging in the region has diminished the supply. South of the taiga region are areas of temperate broadleaf forests and **steppe**, an area of treeless, grassland plains.

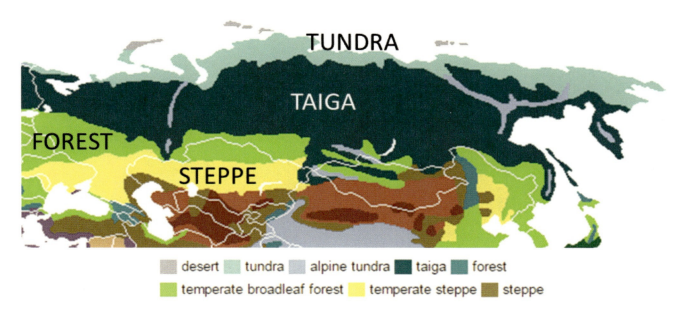

Figure 3.3: Biomes of Russia (Derivative work from original by Sten Porse, Wikimedia Commons)

Although looking at a map, you might assume that Russia has extensive port facilities owing to its vast eastern coastline, it actually has relatively few ice-free ports. Vladivostok, located in the extreme southeastern tip of Russia, is its largest port on the Pacific Ocean (see **Figure 3.4**). Much of the rest of Russia's Far East region is ice-covered throughout the year, making maritime and automotive transport difficult. In fact, this region was only connected to the rest of Russia by highway for the first time in 2010.

Figure 3.4: Port of Vladivostok, Russia (© Dr. Leonid Kozlov, Wikimedia Commons, CC BY-SA 3.0)

Russia's climate more broadly is affected by a number of key factors. In terms of its latitudinal position, meaning its position relative to the equator, Russia is located very far north. In general, as you increase in latitude away from the equator, the climate gets colder. The strong east-west alignment of Russia's major biomes reflects this latitudinal influence. Russia's climate is also affected by its continental position. In general, areas that exhibit a **continental climate** are located near the center of a continent away from water bodies and experience more extremes in temperature due to drier air. Water helps regulate air temperature and can absorb temperature changes better than land. In the winter, areas away from water can be very cold, while in the summer, temperatures are quite hot and there is little precipitation. The third key driver of climate in Russia is its altitudinal position. As you increase in elevation, temperatures decrease. You might have experienced this when hiking in mountains or flying on an aircraft and feeling the cold window. Russia's Ural Mountains, for example, are clearly visible on a map of its biomes as the alpine tundra region owing to its high altitude.

3.2 SETTLEMENT AND DEVELOPMENT CHALLENGES

Russia's size and varied physiographic regions have presented some challenges for its population. Much of Russia is simply too cold for widespread human settlement. Thus, even though Russia is the largest country, the area that is suitable for agriculture and intensive development is much smaller. In Russia's northern regions, agricultural development is restricted by short growing seasons and frequent droughts. As snow melts, it takes topsoil with it, and thus erosion is a serious issue in these areas as well.

Still, some have carved out settlements in this frigid environment. Oymyakon, located in northeastern Russia, is considered to be the coldest permanently inhabited places in the world (see **Figure 3.5**). It has a population of around 500 and temperatures here once dropped to -71.2°C (-96°F)! It takes 20 hours to reach Oymyakon from the nearest city of Yakutsk.

Figure 3.5: Map of Oymyakon, Russia (Derivative work from original by Marmelad, Wikimedia Commons)

Industry, too, is hampered by Russia's cold climate in the Siberian region. Although Siberia accounts for over three-quarters of Russia's land area, it contains only one-quarter of its population. In a region so sparsely populated, how do you build roads, factories, and large settlements? Even if there are resources present, as there are, how do you get them to nearby industrial areas? The industrial developments and human settlements that do exist in this region require high energy consumption and highly specialized facilities needed to cope with cold temperatures and permanently frozen soil.

However, global changes in climate have had some dramatic effects on Russian geography. Areas that were previously permafrost have begun to thaw, leading to erosion and mud, which both present challenges for development. In Siberia, giant holes in the ground began to appear around 2014 and initially baffled scientists. These massive holes were later found to be pockets of methane gas trapped in previously frozen soil that had thawed due to the warming climate. If global temperatures continue to climb, the area of permafrost will shrink, increasing the potential for agriculture in northern Russia. New oil and gas reserves that were previously trapped under frozen soil could likewise become available. Previous shipping routes along Russia's eastern and northern coasts that were covered in ice could become passable.

While warming temperatures might seem beneficial for Russia's frigid northern region, they are accompanied by more troublesome long-term concerns. It is estimated that a huge amount of carbon, around 1600 gigatons (or 1.6 trillion tons), is stored in the world's permafrost. The methane and carbon released from these permafrost stores could exacerbate global warming. Changing temperatures have also been associated with the increased risk of wildfires. In Russia, peatlands, areas of partially decayed vegetation, are particularly at risk. There has been an increase in droughts and flooding throughout Russia and many scientists believe that Russia's close proximity to the Arctic Circle makes it even more vulnerable to changes in temperature.

Climate factors have also shaped the distribution of Russia's population. Most of Russia's population lives west of the Ural Mountains where the climate is more temperate and there are more connections with Eastern Europe (see **Figure 3.6**). Russia is highly urbanized, with almost three-quarters of the population living in cities. Its largest city and capital, Moscow, is home to around 12 million people.

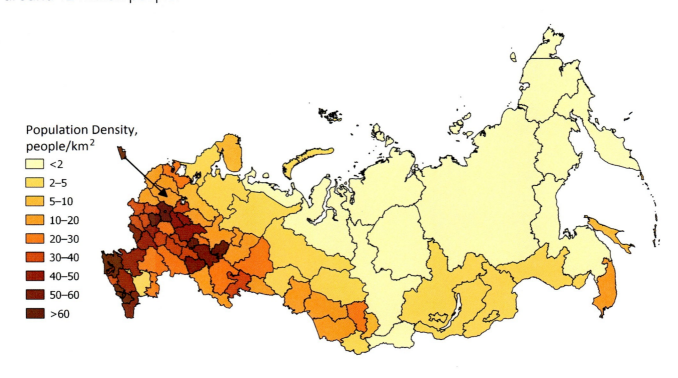

Figure 3.6: Population Density in Russia, 2012 (Derivative work from original by Kartoshka1994, Wikimedia Commons)

Russia's population has experienced some interesting changes over the past few decades. Its population peaked at over 148 million in the early 1990s before experiencing a rapid decline. When geographers explore a country's population, they don't just ask "Where is it changing?" but also *"Why* is it changing?" For Russia, the economic declines coinciding with the dissolution of the Soviet Union contributed to low birth rates. Generally, when a country experiences economic decline or uncertainty, people tend to delay having children. Today, due to higher birth rates and a government push to encourage immigration, Russia's population growth has stabilized and could grow from 143.5 million in 2013 to 146 million by 2050. Russia's death rate remains quite high, however, at 13.1 per 1000 people compared to the European Union average of 9.7 per 1000. Alcoholism rates are high, particularly among men in Russia, and cardiovascular disease accounts for over half of all deaths. In addition, although Russia is highly urbanized, more people are now moving from Russia's crowded cities to more sparsely populated rural areas, in contrast to the more common rural to urban migration seen elsewhere in the world.

3.3 RUSSIAN HISTORY AND EXPANSION

Russia's current geographic landscape has been shaped by physical features, such as climate and topography, as well as historical events. Why is the capital of Russia Moscow, and why is its population so clustered in the west? In the 13th century, Moscow was actually an important principality, or city-state ruled by a monarch. The Grand Duchy of Moscow, or Muscovy as it was known in English, became a powerful state, defeating and surrounding its neighbors and claiming control over a large portion of Rus' territory, an ancient region occupied by a number of East Slavic tribes. The **Slavs** represent the largest Indo-European ethno-linguistic group in Europe and include Poles, Ukrainians, Serbs, as well as Russians.

From the mid-1400s onward, the Muscovite territory expanded at an impressive rate (see **Figure 3.7**). In 1300 CE, the territory occupied an area of around 20,000 square kilometers; by 1462 CE, that number increased to 430,000 square kilometers. By 1584 CE, the territory had swelled to 5.4 million square kilometers.

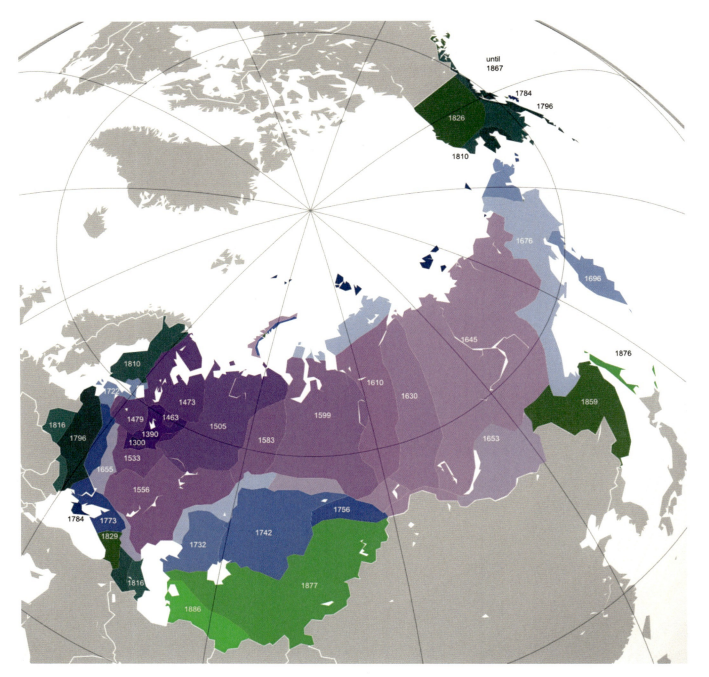

Figure 3.7: Growth of Russia, 1300-1900 (Map by Stephan Steinbach, www.alternativetransport.wordpress.com, CC BY-SA)

During this time, Russia's government shifted as well. In 1547 CE, Grand Duke Ivan IV, better known as "Ivan the Terrible," crowned himself the first Tsar. The term tsar, also spelled czar, stems from the Roman title "Caesar" and was used to designate a ruler, much like the term "king" or "emperor." Ivan IV nearly doubled the territory of Russia during his reign, conquering numerous surrounding ethnic groups and tribes.

Russia's status as an "empire" dates back to the 1700s under the rule of Peter the Great. Peter was able to conquer Russia's northwestern regions, establishing eastern seaports and founding the **forward capital** of Saint Petersburg along the Baltic Sea. A forward capital is a capital that has been intentionally relocated, generally because of economic or strategic reasons, and is often positioned on the edge of contested territory. Overall, the reforms of Peter the Great transformed the country and made it more similar to Western Europe.

The conclusion of World War I coincided with the end of the Russian Empire. The Russian Army fared poorly in the war, with approximately 1.7 million casualties. Russia's people felt that the ruling class had become detached from the problems of everyday people and there were widespread rumors of corruption. Russia went through a rapid period of industrialization, which left many traditional farmers out of work. As people moved to the cities, there was inadequate housing and insufficient jobs. The economic and human cost of World War I, coupled with the plight of workers who felt exploited during the Industrial Revolution, ultimately led to the overthrow of Nicholas II who, along with his family, was imprisoned and later executed.

Eventually, the **Bolsheviks**, a Marxist political party led by Vladimir Lenin, overthrew the interim government and created the Union of Soviet Socialist Republics, abbreviated as the USSR and sometimes simply called the Soviet Union. The capital was also moved back to Moscow from Saint Petersburg. After Lenin's death in 1924, Joseph Stalin took control and instituted both a socialist economy and collective agriculture. Ideally, the changes the Bolsheviks supported were intended to address the failures of Nicholas II providing more stable wages and food supplies. Rather than have individual peasant farms that had limited interconnections and poor systems of distribution, the state would collectivize farming with several farming families collectively owning the land. Under a **command economy**, the production, prices of goods, and wages received by workers is set by the government. In the Soviet Union, the government took control of all industries and invested heavily in the production of capital goods, those that are used to produce other goods, such as machinery and tools. Though this system was intended to address concerns and inequalities that had developed under the tsars, the Soviet government under Stalin was fraught with its own economic and social problems.

3.4 RUSSIAN MULTICULTURALISM AND TENSION

During the period of Russia's expansion and development as an empire, and later during the time of the Soviet Union, Russia's territory included not only ethnic Russians but other surrounding groups as well. **Ethnicity** is a key feature of cultural identity and refers to the identification of a group of people with a common language, ancestry, or cultural history. Many of these minority ethnic groups harbored resentment over being controlled by an imperial power.

Prior to the Bolshevik Revolution, the Russian Empire's response to the non-Russian communities they controlled was known as **Russification**, where non-Russian groups give up their ethnic and linguistic identity and adopt the Russian culture and language. This type of policy is known as cultural **assimilation**, where one cultural group adopts the language and customs of another group. The Russian language was taught in schools and minority languages were banned in public places. Catholic schools were banned and instead, Russian Orthodoxy, part of the **Eastern Orthodox Church**, was taught at state-run schools. The Russian Empire essentially sought to make everyone in the territory Russian. This policy was only marginally successful, however, and was especially difficult to implement in the outer regions.

Under the Soviet Union, the policy of cultural assimilation had less to do with becoming Russian and more to do with being part of the Soviet Union, what could be thought of as Sovietization. The Soviet government organized the country as a federation, where territories within the country had varying degrees of autonomy (see **Figure 3.8**). The larger ethnic groups formed the Soviet Socialist Republics, or SSRs. The Uzbek SSR, for example, largely contained members of the Uzbek ethnic group. The Kazakh SSR similarly consisted mostly of people who were ethnically Kazakh. These SSRs did not represent all of the ethnic diversity present in Russia nor did they provide these territories with autonomy. You might recognize the names of these republics as they gradually became independent states after the collapse of the Soviet Union. The Turkmen SSR became Turkmenistan, for example. Many of these areas, particularly in Central Asia, are majority Muslim and thus adopted the Persian suffix "-stan" meaning "place of" or "country" after independence.

Figure 3.8: Map of the USSR (Courtesy of Perry-Castañeda Library of the University of Texas at Austin, Public Domain)

Under Soviet Rule, some policies of Russification expanded. In Muslim areas of Central Asia and the Caucasus, the use of the Arabic alphabet, the language of the Qur'an, was abolished. The government also sent many Russians into majority non-Russian areas to further unify the country. Other ethnic groups, particularly those perceived as troublemakers by the government, were deported from their ancestral homelands and resettled elsewhere. The ethnic map of the former Soviet Union today, in part, reflects this multicultural history and the legacy of resettlement policies (see **Figure 3.9**). Over 3 million people were deported to Siberia between 1941 and 1949, a large portion of whom died from disease or malnutrition. Others were deported from the Baltic area or from the area near the Black Sea. Overall, around 6 million people were internally displaced as a result of the Soviet Union's resettlement policies and between 1 and 1.5 million of them died as a result.

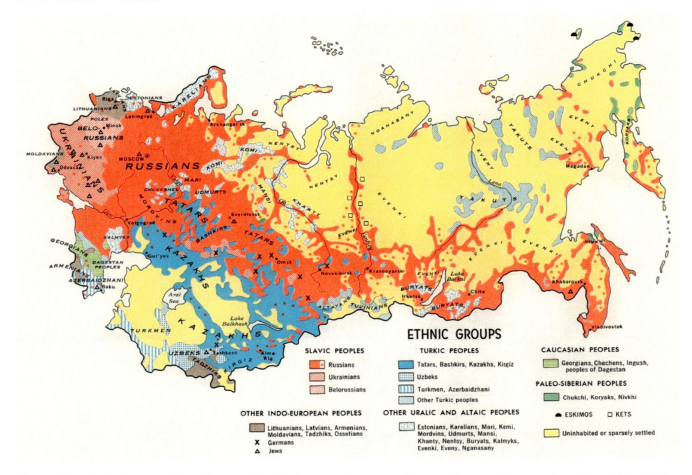

Figure 3.9: Ethnic Groups in the Former USSR (Library of Congress, Geography and Map Division, Public Domain)

Although Russia today is comprised mostly of people who speak Russian and identify with the Russian ethnicity, it contains 185 different ethnic groups speaking over 100 different languages. The largest minority groups in Russia are the Tatars, representing around 4 percent of the population with over 5 million people, and Ukrainians at around 1.4 percent or almost 2 million people. Other ethnic groups, like the Votes near Saint Petersburg, have only a few dozen members remaining. Because of the Soviet resettlement policies, the former Soviet republics have sizable Russian minorities. Kazakhstan and Latvia, for example, are almost one-quarter Russian. This has often led to tension within Russia as minority groups have sought independence and outside of Russia as ethnic groups have clashed over leadership.

In Ukraine in particular, tension between the Ukrainian population and Russian minority has remained high and represents a broader tension between the Eastern European regions that are more closely aligned with Russia and those that seek greater connectivity and trade with Western Europe. Eastern Ukraine is largely comprised of Russian speakers, whileWestern Ukraine predominantly speaks the state language of Ukrainian (see **Figure 3.10**). Overall, around three-quarters of people in Ukraine identify with the Ukrainian ethnicity.

Figure 3.10: Map of the usage of the Russian language in Ukraine, 2003 (© User:Russianname, Wikimedia Commons, CC BY-SA 2.5)

In 2014, the tension between the two groups escalated as then-president Victor Yanukovych backed away from a deal to increase connections with the European Union and instead sought closer ties with Russia. In Western Ukraine, people engaged in widespread protests prompting the government to sign a set of anti-protest laws, while in Eastern Ukraine, most supported the government. Ultimately, Yanukovych was removed from office prompting military intervention from Russia.

Specifically, Russia sought control of Crimea, an area that had been annexed by the Russian Empire and was an Autonomous Soviet Socialist Republic until the 1950s when it was transferred to Ukraine. After the 2014 protests, a majority of the people of Crimea supported joining Russia and it was formally annexed by Russian forces. The region is now controlled by Russia (see **Figure 3.11**). The international community, however, has largely not recognized Crimea's sovereignty or Russia's annexation. This conflict again escalated in 2022 when Russia invaded Ukraine, leading to the largest European refugee crisis since World War II.

Figure 3.11: Map of Crimea (Derivative work from original by Crosswords, Wikimedia Commons)

Several other ethnic groups that remain in Russia desire independence, particularly in the outskirts of the country in the Caucasus region along Russia's border with Georgia and Armenia (see **Figure 3.12**). Chechnya is largely comprised of Chechens, a distinct Sunni Muslim nation. The territory opposed Russian conquest of the region in the 19th century but was forcefully incorporated into the Soviet Union in the early 20th century. 400,000 Chechens were deported by Stalin in the 1940s and more than 100,000 died. Although Chechnya sought independence from Russia, sometimes through violent opposition, it has remained under Russian control following the collapse of the Soviet Union. Dagestan has been the site of several Islamic insurgencies seeking separation from Russia. Ossetia remains divided between a northern portion controlled by Russia and a southern region controlled by Georgia.

Figure 3.12: Map of the Caucasus Region (© Jeroenscommons, Wikimedia Commons, CC BY 2.5)

In an area as large and as ethnically diverse as Russia, controlling the territory in a way that is acceptable to all of its residents has proven difficult. In many large countries, the farther away you get from the capital area and large cities, the more cultural differences you find. Some governments have embraced this cultural difference, creating autonomous regions that function largely independently though remain part of the larger state. Stalin and Russia's tsars before him tried to unify the country through the suppression of ethnic difference, but ethnic and linguistic identities are difficult to obliterate.

3.5 ECONOMICS AND DEVELOPMENT IN THE SOVIET UNION

The Soviet Government, led by Lenin and later by Stalin, advocated a communist system. In a capitalist system, market forces dictate prices according to supply and demand. Those who control the means of production, known as the bourgeoisie in the Marxist philosophy, are much wealthier than the workers, known as the proletariat. In a communist system, however, the means of production are communally owned, and the intended result is that there are no classes of rich and poor and no groups of landowners and landless workers.

In reality, no government practices pure capitalism or pure communism, but rather, governments are situated along a continuum (see **Figure 3.13**). Anarchy, the absence of government control, exists only in temporary situations, such as when a previous government is overthrown and political groups are vying for power. In most Western countries, a mix of capitalism and socialism, where economic and social systems are communally owned, is practiced to varying degrees. Denmark, for example, which has been consistently ranked as one of the happiest countries in the world, has a market economy with few business regulations but government funded universal healthcare, unemployment compensation, and maternity leave, and most higher education is free. The United States is largely capitalist, but the government provides retirement benefits through the social security system, funds the military, and supports the building and maintenance of the interstate highway system. Although China's government is communist, it has also embraced elements of the market economy and allows some private enterprise as well as foreign trade and investment. All governments must address three basic questions of economics: what to produce, how to produce, and for whom to produce. The answers to these questions vary depending on the state and the situation.

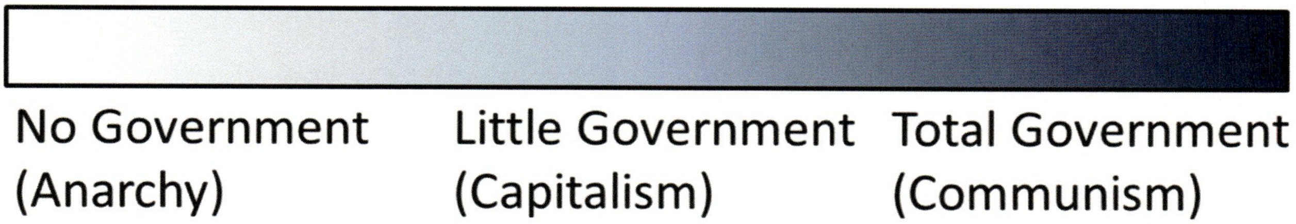

| No Government (Anarchy) | Little Government (Capitalism) | Total Government (Communism) |

Figure 3.13: Continuum of Government Control (Figure by author)

In the Soviet system, the government dictated economic policy, rather than relying on free market mechanisms and the law of supply and demand. This required the government to intervene at all levels of the economy. The prices of goods needed to be set by the central government, the production levels of goods needed to be determined, the coordination of manufacturers and distributors was needed – everything that is traditionally accomplished through private individuals and companies in a capitalist model was the responsibility of the Soviet government.

To coordinate such a wide array of goods and services, long-term planning was needed. The Soviet government instituted a series of five-year plans which established long-term goals and emphasized quotas for the production of goods. This system lacked flexibility, however, and was often inefficient in its production and distribution of goods.

The Soviet government had two principle objectives: first, to accelerate industrialization, and secondly, to collectivize agriculture. The collectivization of agriculture, though intended to increase crop yields and make distribution of food more efficient, was ultimately a failure. By the early 1930s, 90 percent of agricultural land in the Soviet Union had become collectivized, meaning owned by a collection of people rather than individuals. Every element of the production of agriculture, from the tractors to the livestock, was collectivized rather than individually owned. A family could not even have its own vegetable garden. Ideally, under such a system, all farmers would work equally and would share the benefits equally.

Unfortunately, the earnings of collective farmers was typically less than private farmers. This led to a reduction in agricultural output as well as a reduction in the number of livestock. Coupled with a poor harvest in the early 1930s, the country experienced widespread famine and food insecurity. It is estimated that 12 million people died as a result of the collectivization of agriculture.

Soviet industrial development, too, was plagued with inefficiencies. In a typical market economy, particular places specialize in the production of certain goods and the system works out the most efficient method of production and distribution. A furniture maker might locate near a supply of hardwood, for example, to minimize transportation costs. A large factory might locate near a hydroelectric plant to ensure an inexpensive power source. Certain places, due to luck or physical geography, have more resources than others and this can lead to regional imbalances. The Soviet government, however, wanted everything and everyone to be equal. If one region had all of the industrial development, then the people in that region would be disproportionately wealthy and the region would be more vulnerable to attack by an outside force. The government also hoped that the dispersal of industry would force the country to be interconnected. If one area had a steel plant and another had a factory that used steel to produce machines, the two would have to rely on one another and neither would have an advantage. Thus, they aimed to disperse industrial development across the country.

If you were a geographer tasked with finding the best location for a new industry, you'd likely take into account underlying resources, such as the raw materials needed for manufacturing and the energy needed to power the factory. You might think about labor supplies and try to locate the industry near a large labor pool. You might also think about how to get the good to consumers efficiently, and locate near a shipping port or rail line. Rather than take these geographic factors into account, however, the Soviet government sought to disperse industry as much as possible. Industries were located with little regard for the location of labor or raw materials. This meant that inefficiencies were built into the system, and unnecessary transportation costs mounted.

The substantial costs of supporting an inefficient system of industrial development were magnified by the costs needed to fund the **Cold War**. The Cold War occurred following World War II and was a time of political and military tension primarily between the United States and the Soviet Union. Western Europe, which was largely capitalist, was divided from the communist Soviet Union by the so-called **Iron Curtain**, a dividing line between the Soviet Union and its satellite states who aligned with the Warsaw Pact, a collective defense treaty, and Western European countries allied through the North Atlantic Treaty Organization (NATO) (see **Figure 3.14**). The Cold War was so named because it was different from a traditional "hot" war in that it did not involve direct military conflict between the United States and the Soviet Union. It did, however, result in armed conflicts in other parts of the world as well as a massive stockpile of military weaponry.

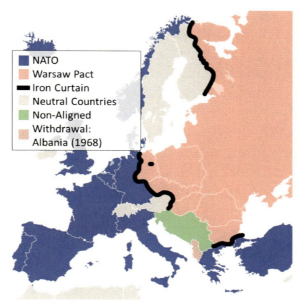

Figure 3.14: Map of Cold War Military Alliances
(Derivative work from original by Sémhur,
Wikimedia Commons)

During the 1980s, Soviet leader Mikhail Gorbachev supported restructuring the Soviet economy with a series of some market-like reforms, known as Perestroika. He also supported glasnost, an increase in government transparency and openness. Unfortunately, these reforms could not change the system quickly enough and loosened government controls only worsened the condition and inefficiencies of the Soviet economy.

The Soviet government, already stretched thin financially from a system of development that largely ignored geography, could not support the unprofitable state-supported enterprises and mounting military expenses. Ultimately, the country went bankrupt. In a system where every aspect of the economy is linked, it only takes one link to break the chain and far-from Gorbachev's policies strengthening the chain, Perestroika only weakened it further. The Soviet Union formally dissolved in 1991. Some have argued that the Soviet Union collapsed economically. Others maintain that it was primarily a political collapse, led by an ineffective government and increasing territorial resistance. Geography largely played a role as well, with the government ignoring fundamental principles of spatial location and interaction.

3.6 THE MODERN RUSSIAN LANDSCAPE

The collapse of the Soviet Union had far-reaching effects on the Russian landscape and even today, Russia is affected by the legacy of the Soviet Union. The remnants of Soviet bureaucracy, for example, affect everything from the cost of road building to the forms needed to get clothes dry cleaned. After the immediate collapse of the Soviet Union, the government transitioned to a market economy. In many cases, those who had positions of power within the Soviet government gained control over previously state-owned industries creating a wealthy class often called a Russian oligarchy. Despite some setbacks and global economic downturns, Russia's economy has improved significantly since the end of the Soviet Union and Russia now has the sixth-largest economy in the world. Poverty and unemployment rates have also fallen sharply in recent decades. Although Russia's population fell sharply following the Soviet Union's collapse, it has rebounded somewhat in recent years.

Abandoned industrial towns and work settlements built by the Soviet Union dot the landscape, evidence of the Soviet government's ill-fated attempt to decentralize its population and development (see **Figure 3.15**). The **Trans-Siberian Railway**, completed in 1916 to connect Moscow with Russia's eastern reaches in Vladivostok, continues to be the most important transportation link in Russia, but Russia's highway system remains largely centralized in the west. In the east, the decentralization of settlements and difficult physical conditions has made building and maintaining road networks difficult. The Lena Highway, for example, nicknamed the "Highway from Hell," is a federal highway running 1,235 km (767 mi) north to south in eastern Siberia. It was just a dirt road until 2014, often turning into an impassible, muddy swamp in summer.

Figure 3.15: Abandoned Apartment Buildings in Kadykchan, Russia (© Laika ac, Wikimedia Commons, CC BY-SA 2.0)

Under Vladimir Putin, Russia's 2nd and 4th president, Russia's economy has grown consistently, aided by high oil prices and global oil demand. Putin also instituted police and military reforms, and persecuted some of the wealthy oligarchs who had taken control of private enterprises. Critics also note that Putin has enacted a number of laws seeking to quiet political dissent and personal freedoms. There have been numerous documented cases of the torture of prisoners and members of the armed forces as well as a number of suspicious killings of journalists and lawmakers.

Although the Cold War officially ended with the collapse of the Soviet Union, tension between Russia and the West remains high. Military conflict in the former Soviet states, like Ukraine, has often reignited simmering hostilities. Still, there is some evidence of cooperation. In 2015, Putin told fellow world leaders that climate change was "one of the gravest challenges humanity is facing" and backed the United Nation's climate change agreement. Previously, Putin had stated that for a country as cold as Russia, global warming would simply mean that Russians would have to buy fewer fur coats. The U.S. and Russian space agencies also continue to work together, announcing plans to cooperatively build a new space station.

CHAPTER 4

North America

> **Learning Objectives**
>
> - Identify the key geographic features of North America
> - Describe how the process of industrialization shaped North American geography
> - Analyze how the patterns of industrialization impacted development in North America
> - Describe the current patterns of inequality in the United States

4.1 NORTH AMERICA'S PHYSICAL SETTING

The giant redwoods that stretch over California's Redwood National Park are the tallest trees on Earth, towering to over 100 meters (328 feet). These trees are also exceptionally old. One such tree, known as "General Sherman," is the largest tree in the world by volume and is believed to be over 2,000 years old. At the time General Sherman first emerged from the ground, North America was settled by a number of indigenous groups. It would be 1,000 more years until Europeans would make contact with the Americas. Today, though many of the redwoods still remain, both the physical and human landscape of North America have profoundly changed.

Traditionally, the continent of North America extends from the Canadian Arctic through the United States and Mexico to the narrow Isthmus of Panama (see **Figure 4.1**). When considering the "region" of North America, however, that is, the area united by common physical and cultural characteristics, there are distinct similarities between Canada and the United States in terms of language and a shared history that are quite different from their Spanish-speaking neighbors to the south. Although the narrow strip of land that typically divides North and South America makes for an easy way to divide these two regions, in many ways, Middle America is largely a transition zone between more powerful economies to the north and south. Mexico, for example, culturally resembles countries like Guatemala and Honduras to the south while physiographically, it resembles the southwestern United States. Thus, the United States and Canada are discussed in the North America chapter while Mexico and Central America are considered alongside the chapter on South America.

Figure 4.1: Map of North America, 1:36,000,000 scale (Derivative work from original by CIA World Factbook, Public Domain)

The physiographic regions of North America are well-defined and are commonly recognized by its residents (see **Figure 4.2**). Someone might say he is from "Appalachia," for example, or that she grew up in the "Rocky Mountains." In general, the physiographic regions have a strong north-south alignment. Climatically, the region is quite diverse, ranging from tundra in northern Canada and Greenland to semi-arid desert in the southwestern United States. These diverse physical conditions have enabled North America to have a wide variety of natural resources, but have also contributed to significant regional differences.

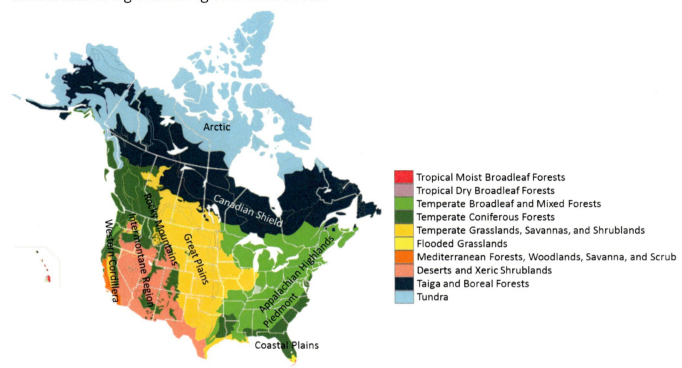

Figure 4.2: Physiographic Regions of North America (Derivative work from original by Cephas, Wikimedia Commons)

Most of Canada's land area consists of **boreal forest**, known as taiga in Russia (see **Figure 4.3**). This boreal forest area consists of coniferous trees, such as spruce and pine, and is characterized by a cold climate. For Canada's indigenous communities in particular, this large stretch of woodland has been an important resource. The rocky landscape of the Canadian Shield extends from the Arctic regions of Central Canada west through Quebec and is among the oldest geologic formations on Earth. It also has some of the world's richest mineral areas.

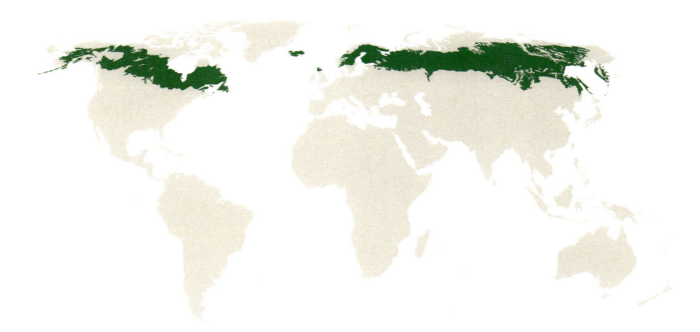

Figure 4.3: Map of Global Boreal Forest Areas (© Mark Baldwin-Smith, Wikimedia Commons, CC BY-SA 3.0)

As with the physical landscape, the climate zones of North America are diverse. In general, North America has a relatively simple weather system. As you increase in latitude north, the temperature decreases and as you travel west to east, the precipitation increases. Thus, California, on the west coast, is relatively warm and dry, while Florida on the east coast is hot and wet.

Most of North America, to include Mexico, Greenland, and some of the Caribbean, is situated on the North American plate and is thus relatively geologically stable (see **Figure 4.4**). One notable exception, however, is the Juan de Fuca Plate, which is subducting under the North American plate near California and Vancouver Island, an area known as the Cascadia subduction zone. Severe earthquakes, generating tsunamis, have occurred here roughly every 500 years; the last major earthquake in the area was in 1700 CE. Just to the south, the San Andreas Fault running along the edge of California forms the boundary between the Pacific Plate and the North American Plate. This is a transform plate boundary, with the two plates sliding past each other horizontally. San Francisco is located on this fault line and the area has experienced numerous earthquakes.

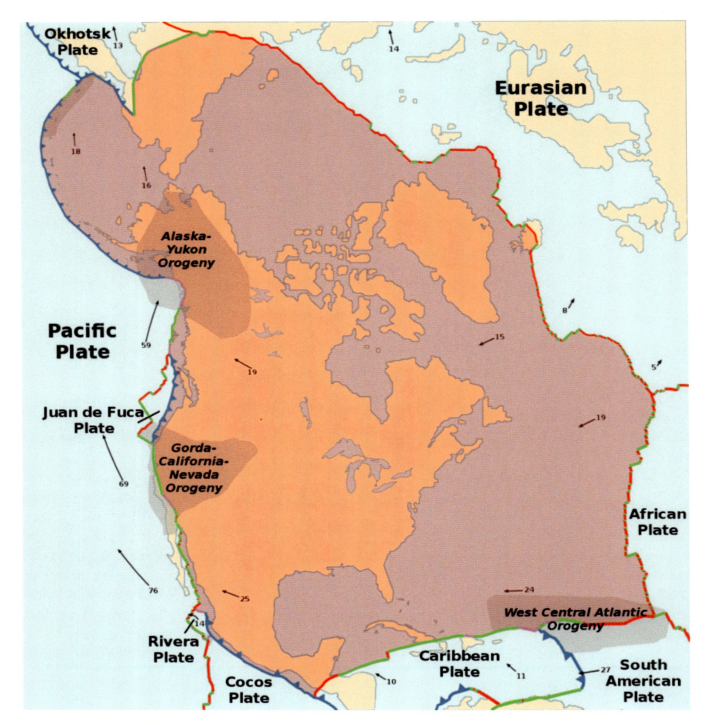

Figure 4.4: Map of the North American Tectonic Plates (Derivative work from original by Alataristarion, Wikimedia Commons)

North America has a number of significant rivers, some of which are used for shipping and others for hydroelectric power. The longest North American river is the Missouri, which forms in Montana and flows into the Mississippi River. The Mississippi River is largely considered to be the most important waterway in terms of commercial transportation. The Port of South Louisiana, located along the Mississippi, is the largest port in the United States in terms of tonnage.

Below North America lies a number of **aquifers**, or underground layers of permeable rock that hold groundwater. The largest of these aquifers is the Ogallala Aquifer located in the central United States stretching from South Dakota down to Texas. This aquifer supplies water to much of the Great Plains – it actually supplies about one-third of all groundwater used for irrigation in the United States. While aquifers are beneficial for irrigation, they replenish their water supplies relatively slowly through rainfall. Depletion of the Ogallala Aquifer has accelerated over the past few decades and currently water is being taken out of the aquifer at a faster rate than it can be replaced. Once all of the water is depleted, it will take around 6,000 years to naturally replenish. Groundwater conservation initiatives in the area have aimed to slow the depletion rate by encouraging farmers to practice sustainable irrigation methods.

While farmers have been encouraged to conserve water, groundwater depletion is just one of the many environmental concerns currently facing North America's farmers. **Sustainable agriculture** more broadly remains an important initiative. This type of agriculture looks at farming's effect on the larger ecosystem and seeks to produce agriculture in a way that doesn't negatively impact the ecosystem in the long-term. It is essentially farming that can be sustained and seeks to minimize water use, soil erosion, and harmful chemicals. Globally, over one-third of all agricultural land has become degraded due to poor land and resource management. Soil is a finite resource, and topsoil can take over 500 years to form! Traditional forms of agriculture, where you might see large stretches of tilled land, can often lead to topsoil erosion. Through sustainable agricultural practices, soil erosion rates have slowed in the United States over the past several decades.

Many environmental problems like topsoil erosion and groundwater depletion affect a wide area and can have far-reaching effects beyond areas where the environment is not being sensitively managed. **Acid rain**, for example, formed from sulfur dioxide and nitrogen oxide emissions, can have damaging effects far beyond the areas that are emitting these gases (see **Figure 4.5**). When cars or factories burn **fossil fuels**, those nonrenewable sources of energy formed by the remains of decayed plants or animals, they release a number of chemicals including sulfur and nitrogen. These gases react with water in the atmosphere to form a highly acidic rain that can damage plants and animals. The lower the pH value, the more acidic a substance is; pure water has a pH of 7. Acid rain can have a pH of around 5.0, or even below 4.0 in some areas. Pickles, by comparison, have a pH of around 5.20, so you can imagine the devastating effects of this acidic precipitation on the environment. The strict regulation of fossil fuel emissions since the 1970s has dramatically reduced instances of acid rain in the United States but some argue that further regulation is needed to address changes in global climate and other pollution concerns.

Map of acid rain severity across the united states

Figure 4.5: Map of Acid Rain in the United States (National Atmospheric Deposition Program/National Trends Network, Public Domain)

4.2 NORTH AMERICAN HISTORY AND SETTLEMENT

Although Christopher Columbus is often credited with "discovering" America, the landmass was inhabited long before Europeans made contact. Most likely, early migrants to the Americas traveled from Asia through the Beringia land bridge that once connected Siberia and Alaska over 10,000 years ago. These indigenous peoples, known as First Nations in Canada or Native Americans in the United States, were divided into a number of different groups, some consisting only of a few small families and others encompassing vast territories and empires (see **Figure 4.6**). Some groups practiced hunting and gathering but many practiced settled agriculture. Before European contact, there were an estimated 50 million indigenous people living in North and South America.

Figure 4.6: North American Indigenous Cultural Areas (Derivative work from original by Spacenut525, Wikimedia Commons)

European colonization completely changed the cultural landscape of North America. In 1492 CE, Columbus made contact with what are now the Bahamas, Cuba, and the island of Hispaniola, spurring Spanish and Portuguese colonization of the Americas. The term "Indian" was actually originally used by Columbus who thought he had arrived in the East Indies, what we now refer to as East and Southeast Asia. Early French and English settlements were not successful, but over time, they too gained control of territory and founded permanent colonies. The easternmost indigenous groups were the first to experience the impacts of European invasion. Many were relocated, often forcibly, to the interior of North America to free up land for European settlement. Disease and war would have a devastating effect on the indigenous groups of the Americas. European settlers and explorers brought smallpox, measles, and cholera – diseases previously unknown to North America. In some areas, 90 percent of the indigenous population died.

By the early 1700s, France, the United Kingdom, and Spain had established formal colonies in the Americas (see **Figure 4.7**) and the population geography of North America today is largely rooted in the colonial developments during this time period. The British primarily set up settlements along the coast, including the thirteen colonies that would declare independence from the United Kingdom and form the basis of the United States. The French colonized much of Canada and the area surrounding the Mississippi River. Their primary objective was fur trading, and they founded a fur trading outpost at what would later become the city of Quebec. The Spanish colonized present-day Florida as well as much of Middle America, stretching into what is now the southwestern United States. They sought resources like gold, the expansion of trade, and opportunities to spread the Roman Catholic faith to indigenous groups.

Figure 4.7: Map of North American Colonies, 1750 (Derivative work from original by Pinpin, Wikimedia Commons)

The early British colonies had highly specialized economies, not unlike the patterns seen in present-day North America. The New England colonies, around the Massachusetts Bay area, were centers of commerce. The Chesapeake Bay area of Virginia and Maryland had a number of tobacco plantations. In the Middle Atlantic, around New York, New Jersey, and eastern Pennsylvania, were a number of small, independent-farmer colonies. Further south, the Carolinas were home to large plantations cultivating crops like cotton.

These large plantations relied on slave labor, a dark legacy that would last for 250 years in North America. Initially, colonists partnered with indentured servants. These laborers paid to their passage to North America by agreeing to work for an employer under contract for a set number of years. These indentured servants often worked on farms, and once their contract expired, they were free to work on their own. Over half of all European immigrants to the Americas before the American Revolution were indentured servants.

As indentured servants gradually earned their freedom, the system of indentured servitude was replaced with slavery. The Portuguese were the first to bring slaves from Africa to the Americas during the 1500s. England, France, Portugal, and the Netherlands would all later join in the transatlantic slave trade, with England dominating the slave trade by the late 17th century. The vast majority of slaves were destined for sugar colonies in the Caribbean and Brazil. Less than 10 percent would be brought to the North American colonies, but this number still represented hundreds of thousands of people. It is estimated that a total of 12.5 million Africans were shipped to the New World as slaves.

During British colonization, slaves worked as house servants or laborers in the northern colonies and farm workers in the south. Britain formally abolished slavery in 1833, but slavery was so entrenched in the economies of the southern United States that it would take a civil war to end the practice. In their secession statement, Mississippi explained its reasoning for leaving the union: "In the momentous step which our State has taken of dissolving its connection with the government of which we so long formed a part, it is but just that we should declare the prominent reasons which have induced our course. Our position is thoroughly identified with the institution of slavery – the greatest material interest of the world. Its labor supplies the product which constitutes by far the largest and most important portions of commerce of the earth" (http://avalon.law.yale.edu/19th_century/csa_missec.asp).

When we think about the Civil War, it is important to understand the geographical differences between the north and south and to remember that the northern states profited on slavery in the south. Just as geographers can divide the world into core and peripheral countries today, the early United States can similarly be analyzed in terms of its core and periphery. The southern states were indeed peripheral in terms of their economic development. Slavery, essentially free labor, provided the southern states with the maximum profit for their commodities and the notion of "othering," the idea that people who look different from you are definitively not you, combined to create an institution that was deeply a part of the southern culture and economy. Even after slavery was abolished in the United States in 1865 with the 13th Amendment to the Constitution, the legacy of slavery and the tendency to consider African Americans as "other" remained. It would be another 100 years before laws were passed in the United States that would bar discrimination based on race, color, religion, sex, or national origin. Even still, racial and ethnic prejudices continue to be a significant social issue.

4.3 INDUSTRIAL DEVELOPMENT IN NORTH AMERICA

As the Industrial Revolution began in the United Kingdom in the mid-1700s and spread across Europe, the United States was still primarily based on agriculture and natural resource production. Some of the early innovations in industry were thus based on these raw resources, such as the cotton mill and textile factories. Hydropower was the key source of energy for these early manufacturing plants and thus they were located almost exclusively in the northeastern United States, the only area with fast-moving rivers. After the Civil War in the 1860s, steam power manufacturing spread through the United States allowing the southern states to industrialize. The manufacturing core region in particular had high concentrations of industrial output (see **Figure 4.8**). Eventually, as the United States continued to industrialize, they overtook the United Kingdom by the early 20th century as the global leader in industry.

Figure 4.8: Map of the North American Manufacturing Core Region (Derivative work from original by Lokal Profil, Wikimedia Commons)

The geography of North America shaped industrial development and regional specializations. In the Pittsburgh-Lake Erie region, for example, abundant deposits of iron fueled steel manufacturing, inspiring the name of Pittsburgh's professional football team. In the south, textile manufacturing developed and remains a regional specialty in many areas still today. Coal from Appalachia fueled industrial development in the Mid-Atlantic States. These regional specializations, and the fact that the southern states continued to rely on agricultural production for some time, further exacerbated economic differences between the north and south.

The Industrial Revolution shaped the pattern of human settlement in North America. As in Europe, industrial development occurred in urban areas spurring people to move from rural farming communities to the cities to find work. In 1790, around 5 percent of the US population lived in urban areas. At the end of the Civil War, as industrialization began to diffuse across the continent, around 20 percent lived in cities. By 1920, more people lived in cities than in rural areas. Today, over 80 percent of people in the US live in cities.

In addition, industrial development spurred large-scale migration, particularly from the peripheral regions of Eastern Europe, as people moved to the US to find work. Between 1865 and 1918, 27.5 million people migrated to the US. Conditions for many of these workers was dismal and child labor wouldn't end until 1930. Asians primarily migrated to the western United States where they were often met with strong anti-immigrant sentiment. Legislation actually limited immigration from China and Japan at the turn of the 20th century. Improvements in rail transportation further diffused both industrial development and the population of workers.

For the past several decades, manufacturing has been declining in the United States as people have shifted to jobs in service industries, like retail and finance. Still, the US remains the world's second largest manufacturer behind China. This process is referred to as **deindustrialization** and is accompanied by both social and economic changes as a country shifts from heavy industry to a more service-oriented economy.

4.4 THE NORTH AMERICAN URBAN LANDSCAPE

North America's urban landscape has been shaped both by colonization and by industrialization. Most of the early settlements in the region were small and were located close to the eastern coast. The Appalachian Mountains provided a formidable obstacle for early settlers before 1765. As settlement and colonization expanded, people moved steadily westward, still primarily situating close to waterways. Even today, most urban centers are located close to water.

During this time, immigration and natural growth expanded North America's population. In 1610, the population of what is now the United States, excluding indigenous groups, was a meager 350 people. In just 200 years, the population reached over 7 million. In 1620, just 60 people occupied what is now the Canadian city of Quebec. Today, the population of the United States stands at over 318 million and Canada's population is over 35 million and both countries are highly urbanized.

North America's cities themselves have also changed over time. The traditional North American city had a core commercial area, called the **central business district** (or CBD), surrounded by worker's homes. Density was generally highest near the city center and decreased as you traveled outward away from the urban center and into the rural areas.

As deindustrialization occurred, suburbanization replaced the previous rural to urban migration. The rush to move to the city center for jobs in industry was replaced by the desire for more land and spacious, single-family homes. With the decrease in housing density and the increase in both home size and acreage, however, came sprawl. **Urban sprawl** refers to the expansion of human settlements away from central cities and into low-density, car-dependent communities. Sprawl is associated with **urban decentralization**, the spreading out of the population that resulted from suburbanization. Counterurbanization, the shift in populations from urban centers to suburban and rural settlements, has been prevalent in North America since the end of World War II. In some areas, rural populations have actually grown as a result of counterurbanization. As sprawl continued, edge cities developed. An edge city is an urban area situated outside of the traditional central business district.

In historical North American cities, the central city was home to most of the jobs and services and had relatively high density housing. Because everything was located close to the city center, people could often walk from home to work or take efficient transit systems like streetcars. Urban decentralization has not only resulted in sprawl but has also created suburbs that are entirely dependent on automobiles (see **Figure 4.9**).

Figure 4.9: Suburban Development in Colorado Springs, Colorado (© David Shankbone, Wikimedia Commons, CC BY-SA 3.0)

Few suburbs have shops or restaurants, and most people living in suburbs have to commute to work. Since jobs are no longer clustered in the city center, cities have faced challenges trying to develop mass transit systems that tie together numerous disconnected suburban developments and link people with their places of work, many of which are now located in surrounding edge cities.

Toronto, for example, Canada's largest city, has a population of 2.8 million within its city limits. Its surrounding suburbs, however, have grown considerably in recent decades. The entire metropolitan area now has a population of over 5.5 million and the average daily commute time is over 1 hour. To the south, Washington, DC's urban decentralization has extended north into Maryland and south into Virginia. Its subway system, a technological marvel when it opened in 1976, has not kept pace with its urban growth and numerous sections of rail lines were shut down for an extended period in 2016 and again in 2019 to conduct major system and station repairs.

In some areas, the metropolitan area has grown so large that it actually overlaps with neighboring metropolitan areas. This is referred to as a **megalopolis**. The Northeast Megalopolis extends along the Interstate 95 corridor from the southern suburbs of Washington, DC north through Baltimore, Philadelphia, and New York to Boston (see **Figure 4.10**). It covers about 2 percent of the land area in the United States but is home to over 50 million people, around 16 percent of the US population. It is projected to grow to 58 million people by 2025. The Northeast Megalopolis is just one of many growing urban areas in North America. The Atlanta Metropolitan area may one day extend into Charlotte, North Carolina. Toronto's urban development may creep south, intermixing with development in Detroit, Cleveland, and Chicago. Florida may one day become one megalopolis linking the cities of Tampa, Orlando, Miami, and Jacksonville. These massive urban settlements will provide new opportunities for creative housing and transportation planning.

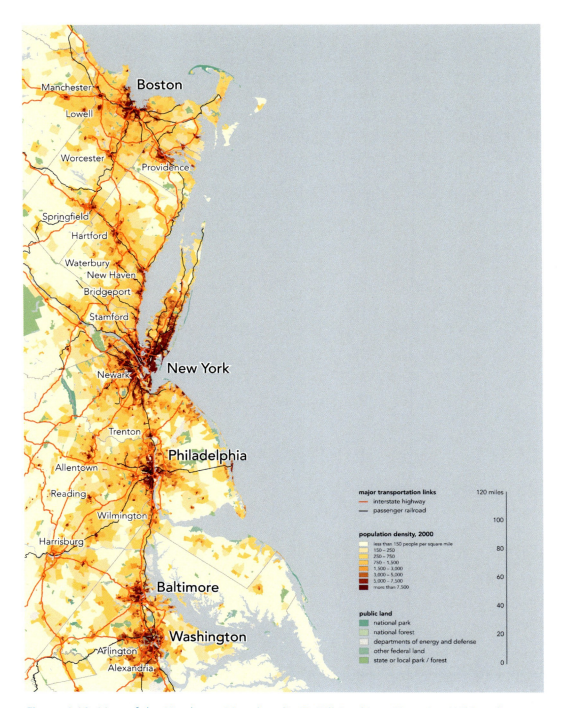

Figure 4.10: Map of the Northeast Megalopolis (© Bill Rankin – Citynoise, Wikimedia Commons, CC BY-SA 3.0)

One creative approach to the problem of urban sprawl is New Urbanism, a movement to create urban landscapes with walkable neighborhoods, accessible public spaces, and housing and shops in close proximity. In the United States alone, more than 600 towns and villages have been developed following the New Urbanist principles. Celebration, Florida, for example, near Orlando was designed and built by the Walt Disney Company and includes a variety of apartments and single-family homes in close proximity to shops, restaurants, and a movie theater – all of which are in walking distance for residents (see **Figure 4.11**). In other areas, New Urbanism is more broadly integrated into long-term urban plans. One criticism of New Urbanist developments is that while on the surface, they promote mixed income developments, in practice most housing in these areas are for the middle and upper classes. Housing prices in these developments are simply beyond the reach of many low income families.

Figure 4.11: Market Street in Celebration, Florida (© Simonhardt93, Wikimedia Commons, CC BY-SA 4.0)

As urban to suburban migration continued, some desired instead to move back from the sprawling suburbs to be closer to the amenities of the downtown area. This often led to **gentrification**, where increased property values displace lower-income families and small businesses. Initially, low-income, historic housing near the city center attracted middle- and upper-income families. As these families moved in and renovated the housing, other families did the same. Over time, this renovation increased property values – an advantage for city officials who saw an increase in property tax revenue. For the poorest in the communities, however, this increase in property values often meant that they could no longer afford to rent near the central city. Given the auto-dependency of the sprawling suburbs, where would someone live if they had no transportation and worked in the downtown area? The walkability of the downtown, an amenity for those relocating from the suburbs, was often a necessity for low-income workers.

Gentrification also changes the racial and ethnic makeup of neighborhoods, as most people moving into these changing urban areas are typically white. The Bedford-Stuyvesant area of Brooklyn, for example, was traditionally an African American community but beginning in the

2000s, began to experience gentrification (see **Figure 4.12**). The percentage of white residents increased from 2.4 percent in 2000 to 22 percent in 2013. Median home prices jumped, too, from $400,000 in 2011 to $765,000 in 2016. New businesses have located in the area and the gentrification has funded major infrastructure improvements. For the neighborhood's poorest residents, however, these improvements have pushed housing and rent prices beyond what they can afford.

Figure 4.12: Gentrified Neighborhood in Bedford-Stuyvesant, Brooklyn, New York (© Mark Hogan, Flickr, CC BY-SA 2.0)

4.5 PATTERNS OF INEQUALITY IN NORTH AMERICA

While both Canada and the United States have relatively strong economies, income inequality persists. In the United States in particular, around 12 percent of people live below the poverty line. Some argue, however, that the traditional definition of "living below the poverty line" has not kept up with rising living costs and inflation and that the actual percentage of Americans living in or near poverty is far higher. This income inequality is geographical, with the states in the south having significantly greater concentrations of people in poverty that the rest of the country (see

Figure 4.13). These regional differences are connected to historical differences in development. Just as the northern areas were the first to industrialize, they were the first areas to transition to more higher-income service industries. Although areas like Silicon Valley in California and the Austin-San Antonio region of Texas have had an influx of high-tech industries, some areas of the south have been slow to transition from primarily agricultural and natural resource based economies.

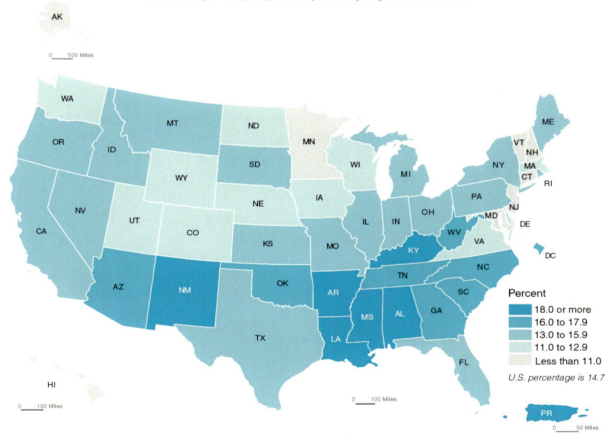

Figure 4.13: Map of Poverty in the United States, 2015 (United States Census Bureau, U.S. Department of Commerce, Public Domain)

Canada's poverty rate is lower than the United States at around 10 percent. In general, Canada has stronger social welfare programs than the US. All provinces of Canada provide universal, publicly funded healthcare, for example, and a monthly income is provided to those in extreme poverty.

However, in both the United States and Canada, income inequality is closely tied to ethnicity and race. For Canada's First Nations, however, poverty and homelessness rates are much higher than the national average. Half of all indigenous children in Canada live in poverty. In some areas, like Manitoba and Saskatchewan, the number is over 60 percent. In the US, the poverty rate among non-Hispanic whites was just over 10 percent in 2014. For black Americans, the poverty rate was 26 percent. By some measures, the US has the highest degree of income inequality among the advanced economies of the world. In Canada, the richest 10 percent own 57.4 percent of the country's wealth. In the United States, the richest 10 percent own over 75 percent of the wealth in the country, the highest of the twenty most developed countries in the world.

4.6 NORTH AMERICA'S GLOBAL CONNECTIONS

North America continues to have a significant role in global trade and influence. Both Canada and the United States are members of the **Group of Seven** (G7), a political forum of the world's leading industrialized countries that also includes France, Germany, Italy, Japan, the United Kingdom, and the European Union. The G7 was formerly known as the Group of Eight, or G8, and included Russia until the country was suspended and later permanently withdrew. Both the US and Canada are also members of the **World Trade Organization** (WTO), an intergovernmental organization that collectively regulates international trade (see **Figure 4.14**).

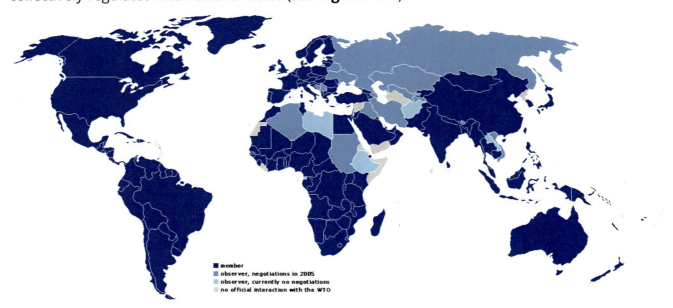

Figure 4.14: Map of the World Trade Organization (© Tsui, Wikimedia Commons, CC BY-SA 3.0)

Within North America, trade has been governed under the **North American Free Trade Agreement** (NAFTA) between Canada, Mexico, and the United States. This agreement was established in 1994 with the goal of increasing economic cooperation between the three countries. Prior to NAFTA, although the US and Canada engaged in free trade, goods bought and sold between Mexico and the US were subject to tariffs, or additional taxes. In 2018, NAFTA was replaced by the United States–Mexico–Canada Agreement (USMCA) as a result of a renegotiation of NAFTA sought by US President Donald Trump.

NAFTA has had generally positive impacts on the economies of the region. Canada's manufacturing output held steady despite global decreases in productivity. Mexico's **maquiladoras**, manufacturing plants that take components of products and assemble them for export, have become a fixture of its landscape especially along the border. The United States also saw a modest economic boost from the agreement.

After the Cold War, the United States retained its position as a global superpower. It has the largest economy of any other country, including the combined output of the European Union, accounting for 25 percent of the world's gross domestic product (GDP). It leads the world in military expenditures, and by many measures, is the most influential country in the world. However, it also has the largest prison population and has a much higher infant mortality rate than most other industrialized countries with strong regional concentrations of high infant mortality. Some wonder if the US will retain its global dominance in the coming decades, or if it will become one country among many influential world leaders.

Both Canada and the United States continue to attract immigrants, drawn to these countries by the hope of good jobs and political freedoms. Each country has dealt with the influx of immigration in very different ways. Over 200,000 people immigrate to Canada every year and the Canadian immigration system gives preference to immigrants for skilled professions. Around 20 percent of Canada's population is foreign born, the highest of the G7 countries. Canada's immigrants have shaped its cultural landscape and have created a rich cultural mosaic. In contrast, immigrants to the United States have generally been expected to assimilate, creating a relatively homogeneous cultural landscape rather than retaining individual ethnic identities. This notion of mixing cultural groups to create a more homogeneous national culture is metaphorically termed a **melting pot**.

Canada and the United States' reactions to refugees have also been markedly different. The United States set a goal of accepting 10,000 Syrian refugees, but immigration from Syria has been contentious politically with some fearing the potential for terrorist attacks by migrants. Several state governors outright refused to accept Syrian refugees. The Canadian government, in contrast, agreed to resettle 25,000 Syrians in 2016. Canadian Prime Minister Justin Trudeau greeted the first plane of refugees, offering winter clothing and stuffed animals and saying, "Welcome home." Throughout history, Canada has welcomed the world's displaced peoples, accepting 1.2 million refugees since World War II.

Undocumented, or illegal, immigration to the United States continues to be another significant political issue. Around 11 million undocumented migrants currently live in the US. Just over 50 percent are from Mexico. As drug crime worsened in Central America, undocumented migration from those countries surged and many now make a long and dangerous trek from Central America through Mexico in hope of reaching US soil. Undocumented and unaccompanied child migrants in particular have increased dramatically in recent years. As countries experience economic decline, political turmoil, and often dangerous living conditions, migrants will likely continue to flock to Canada and the US in search of a better life.

CHAPTER 5

Middle and South America

> **Learning Objectives**
>
> - Identify the key geographic features of Middle and South America
> - Describe the primary patterns of colonial development found in Middle America
> - Analyze the patterns of urban development in South America
> - Explain how globalization has shaped current issues of inequality across Middle and South America

5.1 THE GEOGRAPHIC FEATURES OF MIDDLE AND SOUTH AMERICA

Middle and South America (see **Figure 5.1**) cover an area of the world that is fragmented both in terms of its physical connectivity and its history. Generally, the continents of North American and South America are divided at the **Isthmus** of Panama, the narrow strip of land that connects the two large landmasses. Culturally, though, Middle America, including the Caribbean, is quite similar to South America and this region shares a distinct pattern of colonial development.

MIDDLE AND SOUTH AMERICA

Figure 5.1: Map of Middle and South America (CIA World Factbook, Public Domain)

"Middle America" is typically defined as the area between North and South America, with Mexico sometimes categorized as North America and sometimes as Middle or Central America. Since Mexico shares strong cultural and historical similarities with the countries of Central America, they are grouped together in this text. This region also includes the islands of the Caribbean. South of Middle America is the continent of South America, extending from the the tropical sand beaches of Colombia to the frigid islands of southern Chile and Argentina.

The region lies at the intersection of a number of tectonic plates making the region vulnerable to earthquakes and volcanoes (see **Figure 5.2**). Haiti, for example, located on the eastern half of the island of Hispaniola, is situated on the edge of the Caribbean plate along a transform plate boundary. A magnitude 7.0 earthquake struck here in 2010 killing over 100,000 people.

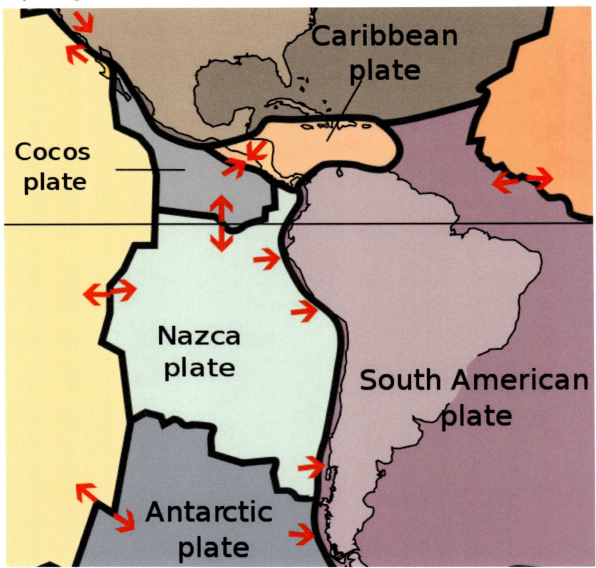

Figure 5.2: Tectonic Plate Boundaries in Middle and South America (CIA World Factbook, Public Domain)

These tectonic collisions have created a landscape of relatively high relief, particularly in Middle America and western South America. Mexico is home to a number of impressive mountain ranges, including the Sierra Madre Occidental in the west, Sierra Madre Oriental in the east, and Sierra Madre del Sur in the south (see **Figure 5.3**).

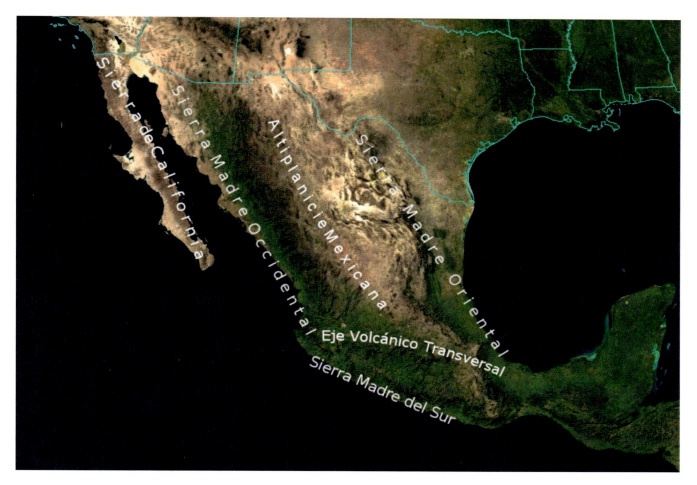

Figure 5.3: Sierra Madre Ranges in Mexico (Satellite image by NASA, Derivative by Ricraider, Wikimedia Commons, Public Domain)

Further east, the tectonic collision of the Caribbean plate and the North American plate formed the Caribbean **archipelago**, or island chain. Many of these islands are the tops of underwater mountains. The islands of the Caribbean are divided into the Greater Antilles and the Lesser Antilles (see **Figure 5.4**). The Greater Antilles include the larger islands of Cuba, Jamaica, Hispaniola, and the Cayman Islands. The Lesser Antilles are much smaller and include the Leeward and Windward Islands, the Leeward Antilles, and the Bahamas.

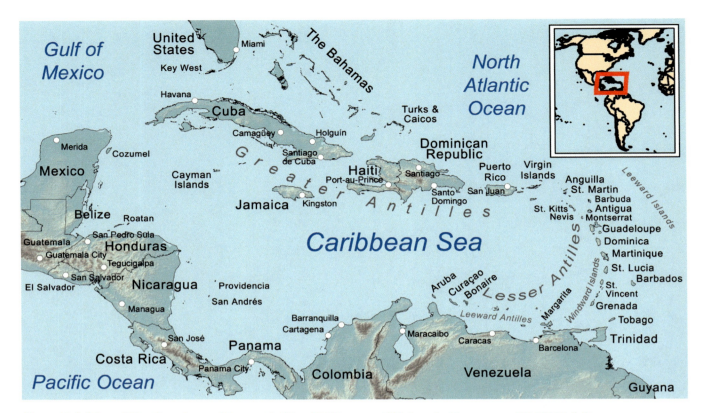

Figure 5.4: Map of the Greater and Lesser Antilles (© Kmusser, Wikimedia Commons, CC BY-SA 3.0)

In addition to earthquakes and volcanoes, the region is prone to tropical cyclones, also known as hurricanes. The areas along the Gulf of Mexico, in particular, lie in the path of frequent hurricanes. **El Niño**, the warming phase of the El Niño-Southern Oscillation (ENSO) cycle, also contributes to severe weather in the region. In North America, El Niño results in warmer than average temperatures, but it can also increase the number of tropical cyclones in the Americas and excessive rain across South America.

The high relief of Central America has created distinct agricultural and livestock zones, known as **altitudinal zonation**. As altitude increases, temperature decreases, and thus each altitudinal zone can support different crops and animals (see **Figure 5.5**). The hot, coastal area known as the *tierra caliente*, for example, can support tropical crops like bananas and rice. Past the tree line in the higher elevation of the *tierra helada*, animals like llamas can graze on cool grasses. In this way, even countries with a relatively small land area can support a wide variety of agricultural activities.

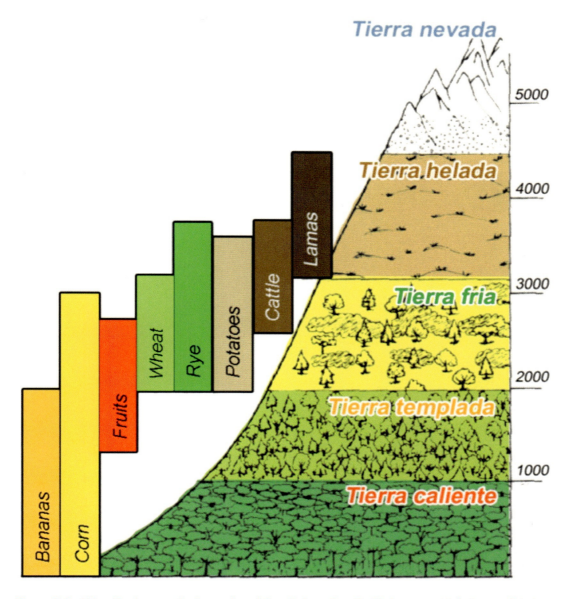

Figure 5.5: Altitudinal zones in Central and South America (© Chris.urs-o, Maksim, and Anita Graser; Wikimedia Commons, CC BY-SA 3.0)

South America's Andes Mountains, which stretch from Venezuela down to Chile and Argentina, were formed from the subduction of the Nazca and Antarctic plates below the South American plate. They are the highest mountains outside of Asia. Situated in the Andes is the **Altiplano**, a series of high elevation plains. These wide basins were central to early human settlement of the continent.

The rest of South America is relatively flat. The Amazon basin is the other key geographic feature of the continent (see **Figure 5.6**). The Amazon River is South America's longest river and is the largest river in the world in terms of discharge. The river discharges 209,000 cubic meters (7.4 million cubic feet) every second – more than the discharge of the next seven largest rivers combined! Its drainage basin covers an area of over 7 million square kilometers (2.7 million square miles).

Figure 5.6: Amazon River and Drainage Basin (©Kmusser, Wikimedia Commons, CC BY-SA 3.0)

Numerous rivers snake across Central America, but the most prominent water feature is Lake Nicaragua. This large, freshwater lake is home to numerous species of fish and provides both economic and recreational benefits to the people of Nicaragua. Plans were approved to build a canal through Nicaragua to connect the Caribbean Sea to Lake Nicaragua and the Pacific Ocean though many worry about the ecological impacts of such a large-scale project.

Currently, the only connection between the Caribbean Sea and the Pacific Ocean is through the Panama Canal. The canal was started in 1881 by the French, in what was then territory owned by Colombia, but the project was a failure. French construction workers were unprepared for the torrential Central American rainy season, dense jungle, and difficult geology. 22,000 workers were killed due to disease and accidents.

The United States helped Panama achieve independence from Colombia and in exchange, Panama granted the US rights to build and control the canal. In 1904, the US continued where the French had left off and in just ten years, the project was completed. Over 5,600 workers died during the US construction project. Panama regained control of the canal in 1999.

When the canal was completed, it accommodated around 1,000 ships per year. Today, around 15,000 ships pass through the canal each year. One key issue is that the Panama Canal's waterways are almost entirely man-made and pass through areas of changing elevation. A series of locks take ships from lower-elevation waterways and raise or lower them depending on the direction the ships are headed. It takes around 8 to 10 hours for a ship to pass through. These locks were not built with modern ships in mind, however, and construction was undertaken to widen the locks to accommodate today's massive container ships. The expansion project was completed in 2016.

5.2 COLONIZATION AND CONQUEST IN MIDDLE AMERICA

Middle America was settled by a number of indigenous groups who originally migrated to the region from North America. Some continued on through the Isthmus of Panama to South America. Here, they founded the Mesoamerican cultural hearth, considered one of the earliest civilizations in the world. Two groups in the region had a particularly strong impact on the cultural landscape of Middle America: the Maya and the Aztec.

The Maya Civilization began around 2000 BCE and stretched across present-day Honduras, Guatemala, Belize, and the Yucatan peninsula. The civilization had a theocratic structure, with their king viewed as a divine ruler. They developed a system of hieroglyphic script, a calendar, a system of mathematics, and astronomy. The civilization had a number of city-states linked by a complex trading system. They also had monumental architecture, and a number of their buildings, like the pyramidal Chichen Itza, are still visible on the landscape today (see **Figure 5.7**). At its height, the Maya Empire encompassed over one million people.

Figure 5.7: El Castillo at Chichen Itza (© Daniel Schwen, Wikimedia Commons, CC BY-SA 4.0)

So what happened to the Maya? The short answer is: we're not really sure. It takes a carefully managed infrastructure to care for one million people. Some researchers think the civilization simply got too big too fast and any number of calamities, perhaps ecological damage or a disease epidemic, related to this rapid population growth could have had a devastating effect. Others think that infighting broke out within the society. Still others maintain that historical climate data shows a decrease in rainfall in the region around the time of the Maya's decline, perhaps indicating a widespread famine. In any case, in such a large society, it would only take a small problem to send the system into turmoil and by the 9th century CE, the Maya had abandoned its cities and its empire collapsed.

The Aztec Empire developed much later in history, during the 15th century CE (see **Figure 5.8**). This civilization was centered around Tenochtitlan which became its capital and one of the greatest cities in the Americas, with a population between 100,000 and 200,000 people. Today, the ruins of Tenochtitlan are located under present-day Mexico City. Aztec architecture, art, and trading systems were truly extraordinary for their time.

Figure 5.8: Map of the Aztec Empire, 1519 (© Badseed, historicair, and Madman2001; Wikimedia Commons, CC BY-SA 3.0)

This empire was relatively short-lived, however, and the reason for its decline is much easier to pinpoint than the Maya. The Spanish, led by the conquistador Hernán Cortés, aligned themselves with a rival group and arrived in Tenochtitlan, as very unwelcome visitors, in 1520. Violence erupted and the Aztec leader Montezuma was killed, marking the beginning of the end of the civilization. By 1521, the Spanish and their allies had destroyed the city of Tenochtitlan and the Aztecs were subsequently ruled by a series of leaders who were chosen by the Spanish.

Colonization completely reshaped the Middle American landscape, from architecture to politics to land-holding patterns. Middle America can be divided into two different spheres, the mainland and the rimland, each with a distinct colonial history and experience (see **Figure 5.9**). The rimland, though a fragmented realm of islands, was more accessible for European colonists than the mainland and were among the first places explorers landed when they reached the Americas. Christopher Columbus first reached the rimland in 1492 CE and would reach South America on a third voyage in 1498 CE. The first Spanish cities in the Americas were established in the region during this time period. By the 1600s, England, Portugal, France, and the Netherlands were shipping Africans to the Americas to work on farms. Of the more than 11 million Africans who were sold as slaves and shipped overseas, over 90 percent were sent to the Caribbean and South America.

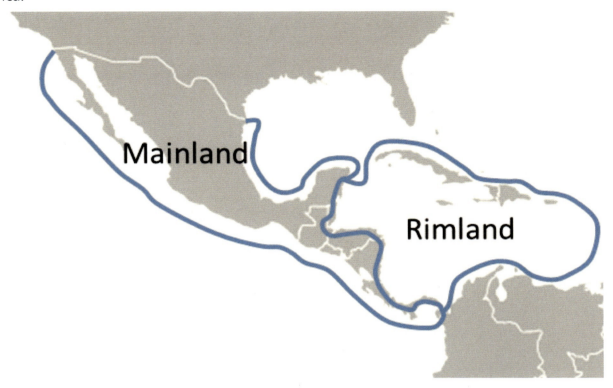

Figure 5.9: Map of the Middle American Mainland and Rimland (Derivative work from original by Lokal_Profil, Wikimedia Commons)

The rimland's sprawling **plantations** where the slaves worked were focused on growing crops, most often sugar, for export. Prior to the commercial agriculture practiced on plantations, most in the region practiced **subsistence farming**, where farmers grow enough food to feed themselves and their families. Since subsistence farmers eat what they grow, they typically grow a variety of crops. You can only eat so much corn before you need a bit more variety. Plantation farms, however, were generally monocultures, meaning only a single crop was grown. This, combined with the free slave labor, allowed for maximum efficiency and profit. Labor was seasonal, coinciding with the seasonality of the cultivated crop. Today, the rimland is still home to a number of plantations and the blending of European and African cultures is prominent on the landscape.

In the mainland, there is a blending of both indigenous and Spanish cultures. There is an ethnic blending as well. **Mestizo** refers to someone of mixed European and Amerindian, or indigenous American, descent and a number of Middle and South American countries have a sizable mestizo population. The Spanish conquest of mainland Middle America not only toppled the Aztec civilization but also led to the deaths of millions of indigenous people due to war and disease.

In contrast with the plantations of the rimland, the mainland was more commonly home to **haciendas**, Spanish estates where a variety of crops were grown both for local and international markets. Because of this variety, workers lived on the land, unlike the seasonal laborers needed to work plantations. Furthermore, although haciendas were less efficient than plantations, they were also less vulnerable economically. If you're only growing one crop and the price of that crop declines, you have no backup. Similarly, disease or a bad harvest could dramatically decrease profits. The increased diversity in crops grown on a hacienda decreased risk. Haciendas also had an element of social prestige. The size of the hacienda increased the social standing of the landowner, and hacienda farmers were often given their own plots of land to cultivate.

Broadly, the colonization of Middle America led to **land alienation**, where land is taken from one group and claimed by another. Where indigenous groups might have previously controlled their own subsistence farms, wealthy European settlers took over the land and built haciendas – often then employing those whose land they claimed. Today, poverty continues to be a significant issue among the indigenous people of Middle and South America and the current system of land ownership found in the region directly connects to European colonization.

5.3 THE SOUTH AMERICAN COLONIAL LANDSCAPE

A variety of ancient cultures were found in South America prior to colonization. These indigenous groups settled in a variety of environments, some in the coastal plains and others in the Amazon basin. One group, the Inca, primarily settled in the altiplano of Peru beginning in the 13th century. The Inca Empire was the largest of the pre-Colombian, referring to before Columbus' arrival, civilizations. Initially, the Inca founded the city-state Kingdom of Cusco, but over time, expanded to encompass four territories stretching 2,500 miles and included over 4 million people.

In 1494 CE, Spain and Portugal signed the Treaty of Tordesillas, dividing up territory in the New World between the two colonial empires (see **Figure 5.10**). The Spanish would control territory to the west of the line while Portugal would control territory to the east. Spanish conquistador Francisco Pizarro reached the Inca by 1526 CE. The empire, already weakened by smallpox and infighting, was conquered by the Spanish soon after. Portugal meanwhile conquered much of eastern South America in present-day Brazil.

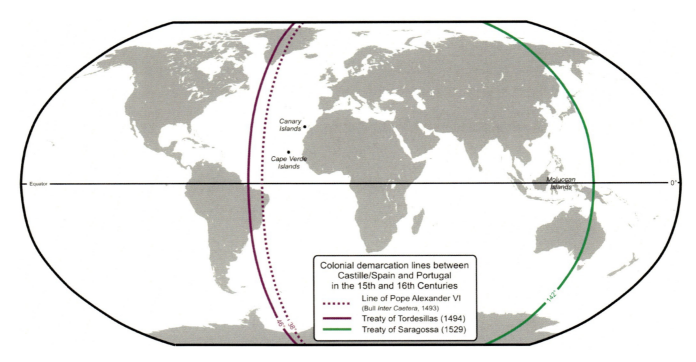

Figure 5.10: Map of the Lines Demarcated by the Treaty of Tordesillas (© Lencer, Wikimedia Commons, CC BY-SA 3.0)

In coastal South America, France, the Netherlands, and the United Kingdom established colonies (see **Figure 5.11**). These colonial possessions were largely extensions of the Central American rimland with large plantations and slave labor. Portugal, too, established plantations along coastal Brazil. As colonialism expanded, the colonial empires prospered. Lima, for example, in present-day Peru, became one of the wealthiest cities in the world due to its silver deposits.

Figure 5.11: Colonies of South America, 1796 (Derivative work from original by Esemono, Wikimedia Commons)

Colonization dramatically changed the urban landscape of the Americas as well as rural development patterns. **Development** broadly refers to economic, social, and institutional advancements and levels of development vary widely across the region. In the rural areas of South America, land was taken from indigenous groups, as it had been in Middle America, and transformed to the benefit of colonial interests. The main interest of the conquering group was to extract riches with little thought given to fostering local development and regional connectivity. Even today, many of the rural areas of South America remain highly isolated and the indigenous descendants of conquered Amerindian groups among the poorest in the region.

In cities that were conquered, European colonizers typically razed existing structures and built new ones. In general, there was little regard for local development and cultural values. The Spanish colonies, for example, were governed according to the Laws of the Indies. These laws regulated social, economic, and political life in territories that were controlled by Spain. They also prescribed a very specific set of urban planning guidelines, including building towns around a Plaza Mayor (main square) and creating a road network on a grid system. Even today, the cities of the Americas often look quite European. In Mexico City, for example, the Spanish destroyed the Aztec capital of Tenochtitlan and built the Mexico City Cathedral over the ruins of the Aztec Templo Mayor complex (see **Figure 5.12**).

Figure 5.12: Mexico City Metropolitan Cathedral, Mexico (© Jeff Kramer, Flickr, CC BY 2.0)

In the early 19th century, most of the colonies of Middle and South America gained their independence, often led by the Europeans who had settled in the region. Larger colonial possessions often separated into smaller independent states. For a short time, the states of Central America formed a federal republic, but this experiment devolved into civil war. Today, most of the mainland of Middle and South America is independent, with the exception of French Guiana which is maintained as a French territory and is home to a launch site for the European Space Agency. Many of the island nations are still controlled by other countries. France, the United States, the United Kingdom, and the Netherlands all still have territories in the Caribbean.

5.4 URBAN DEVELOPMENT IN SOUTH AMERICA

South America is a highly urbanized region, with over 80 percent of people living in cities. Central America and the Caribbean are slightly less urbanized at around 70 percent. Development and human settlement are not spread evenly across the region, however. Several countries in Middle and South America have a **primate city**. Primate cities are those which are the largest city in a country, are more than twice as large as the next largest city, and are representative of the national culture. For example, of Uruguay's 3.4 million people, over half live in its capital and primate city of Montevideo. Not all countries of the world have a primate city. Germany's largest city is Berlin, which is roughly twice as large as Hamburg and Munich and was once the country's primate city. In recent years, however, Munich has increasingly become Germany's cultural center.

The region is also home to several megacities. A **megacity** is a metropolitan area with over 10 million people. Mexico City, the capital and primate city of Mexico, has a population of 22 million people. São Paulo, Brazil has 21.5 million. Rio de Janeiro, Brazil and Buenos Aires, Argentina are also megacities. Megacities often face distinct challenges. With over 10 million people comes a significant need for affordable housing and employment. Megacities often have large populations of homeless people and, particularly in Middle and South America, sprawling slums. This immense population also needs a carefully managed infrastructure, everything from sanitation to transportation, and the developing countries of this region have historically had difficulty meeting the demand.

Despite the challenges faced by large urban populations, rural to urban migration continues in the region. As in many parts of the world, poor rural farmers migrated to the cities where industrial development was clustered in search of work.

In general, the cities of Middle and South America follow a similar model of urban development (see **Figure 5.13**). The central business district, or CBD, is located in the center of the city often alongside a central market. While some colonial buildings were demolished following independence, cities in this region still typically have a large plaza area in the CBD. As industrialization occurred, additional industrial and commercial development extended along the spine, which might be a major boulevard. The spine is often connected to major retail area or mall.

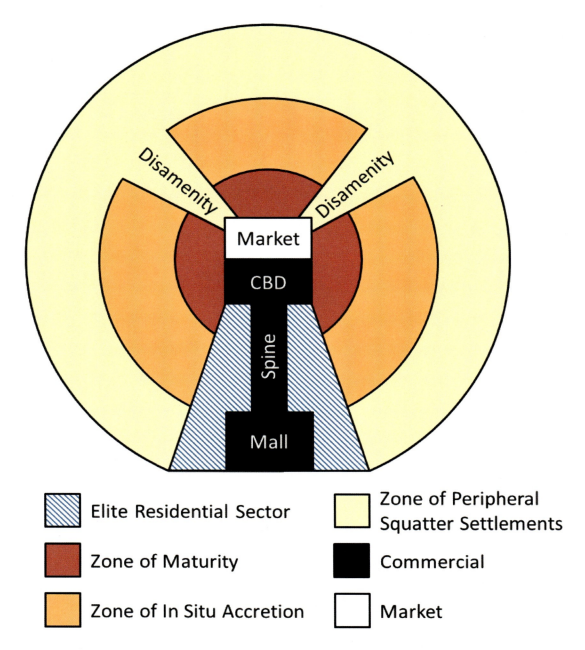

Figure 5.13: Model of the Latin American City (Figure by author, based on Ford 1996)

Surrounding the commercial area of the spine is the elite residential sector consisting of housing for the wealthiest residents of the city often in high-rise condominiums. Around the CBD is the zone of maturity, an area of middle class housing. The zone of in situ accretion is a transitional area from the modest middle class housing of the zone of maturity to the slums of the city's poorest residents.

The outermost ring in a typical Latin American city is the zone of peripheral **squatter settlements**. In this zone, residents do not own or pay rent and instead occupy otherwise unused land, known as "squatting." In some cases, residents in this zone earn money by participating in the **informal sector** where goods and services are bought and sold without being taxed or monitored by the government. Disamenity sectors arise along highways, rail lines, or other small tracts of unoccupied land where the city's poor often live out in the open. Residents often build housing out of whatever materials they can find such as cardboard or tin. What is perhaps most striking about the Latin American city is that in some areas, the city's poorest residents live in an area adjacent to the wealthiest residents magnifying the income inequality that is present in the region.

Globally, around one-third of people in developing countries live in slums, characterized by locations with substandard housing and infrastructure. Estimates vary regarding the total number of people who live in slums but it is likely just below 1 billion people and continues to climb. In Brazil, these sprawling slums are known as favelas and over 11 million people in this country alone live in favelas. Rocinha, located in Rio de Janeiro, is Brazil's largest favela and is home to almost 70,000 people (see **Figure 5.14**). It has transitioned from a squatter area with temporary housing to more permanent structures with basic sanitation, electricity, and plumbing.

Figure 5.14: Rocinha Favela in Rio de Janeiro, Brazil (© Chensiyuan, Wikimedia Commons, CC BY-SA 4.0)

In some cases, those who live in the slums of Middle and South America are not unemployed but simply cannot find affordable housing in the cities. Rural to urban migration here, as in other parts of the world, has outpaced housing construction. Even some lower and middle managers are unable to find housing and thus end up living in the slums.

5.5 INCOME INEQUALITY IN MIDDLE AND SOUTH AMERICA

Although income inequality in Middle and South America has fallen in recent years, this region remains by some measures the most unequal region in the world. Overall, the top 10 percent of people in Latin America control around 71 percent of the region's wealth. If current trends continue, the top 1 percent will have amassed more wealth than the bottom 99 percent. In Mexico, around half of the population lives in poverty and while the rich in Mexico have seen their wealth climb dramatically in recent years, poverty rates remain relatively unchanged. In Brazil, the wealthiest 10 percent of the population own almost three-quarters of the country's wealth, around the same as in the United States. This inequality has been a product of geography but has also impacted the landscape, as well.

Farmers in Middle and South America have struggled with land ownership after their alienation from the land during colonization. While countries like Spain and Portugal no longer control land in Middle and South America, many of these countries' governments took over colonial landholdings during independence rather than turning it back over to private farmers. Often, small farmers in the region simply can't compete with the large-scale agricultural producers. This either worsens rural poverty or contributes to rural to urban migration as farmers leave to find work elsewhere.

Government responses to income inequality vary. Some countries of Latin America and the Caribbean turned to socialism in the hope that government-controlled development would be able to more fairly distribute wealth. Often these socialist endeavors were financed with the exports of natural resources, such as oil or coffee, but this created a vulnerable dependency on foreign trade. In Venezuela, for example, where Hugo Chavez ushered in a socialist revolution at the turn of the 21st century, falling oil prices in 2016 threw the economy into steep decline leading to massive inflation and a shortage of domestic products (see **Figure 5.15**). Governments like Venezuela often relied too heavily on income from exports and invested little in developing their own infrastructure, instead simply relying on importing the goods they needed. In general, spending on social services remains relatively low across the region.

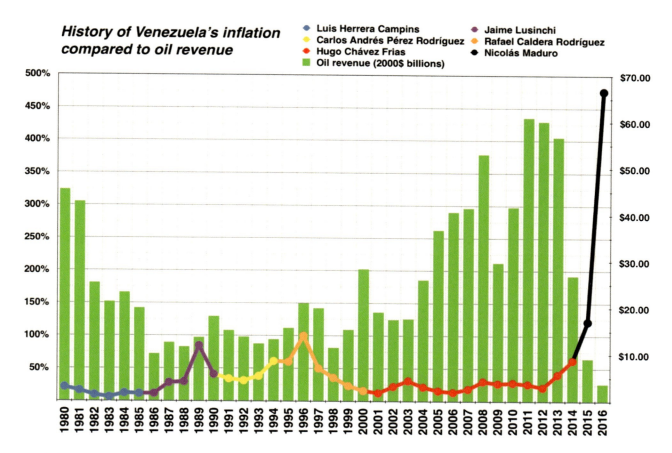

Figure 5.15: Venezuela's Inflation Rate Compared to Annual Oil Revenues, 1980-2015 (© ZiaLater, Wikimedia Commons, CC BY 1.0)

Taxation systems have had a relatively minimal effect on bettering the lives of the region's poor or assisting the region in infrastructure development. The wealthiest people in many of these countries hold their money offshore in order to avoid taxation, but this also prevents governments from being able to use this tax revenue. Many governments have also given tax breaks to large, multinational corporations who seek to do business in the region providing a short-term economic increase at the expense of long-term development planning.

Inequality is not just an issue of poverty, however. It can also relate to unequal access to education and political power. 62 percent of the population of Bolivia is indigenous, for example, but the country did not have a president from indigenous descent until Evo Morales was elected in 1998. Among the indigenous population of Bolivia, most work in agriculture and around 42 percent of indigenous students do not finish school compared to just 17 percent of non-indigenous students. There is a distinct cycle between education and poverty with educational advancement directly linked to economic advancement. In some areas, access to adequate education, particularly among indigenous populations, remains low, limiting the opportunity to narrow the income gap.

For some, **liberation theology** has provided a sense of hope. Liberation theology is a form of Christianity that is blended with political activism. There is a strong emphasis on social justice, poverty, and human rights. This approach also stresses the importance of alleviating poverty through action and followers believe that, like Jesus, they should align themselves with society's marginalized groups.

Others in the region have decided to look elsewhere for economic advancement. Most countries in Middle and South America have net out-migration, meaning more people are leaving than coming into the country. Around 15 percent of all international migrants are from Latin America and the United States continues to be top destination. Some from Central America, however, are choosing to stay in Mexico rather than continue the journey north to the United States.

5.6 PATTERNS OF GLOBALIZATION IN MIDDLE AND SOUTH AMERICA

The ongoing migration from Middle and South America points to a larger issue of global economic connectivity. Many of the region's migrants are well-educated and leave in search of better economic opportunities. This contributes to **brain drain**, referring to the emigration of highly skilled workers "draining" their home country of their knowledge and skills. Around 84 percent of Haiti's college graduates live outside of their home country, for example, the greatest percentage of any country in the world.

When workers leave the region in search of work elsewhere, they often send home remittances, or transfers of money back to their home country. In 2014, global remittances totaled $583 billion, and in some countries, remittances represent a significant portion of the country's GDP, in some cases exceeding the amount the country earns from its largest export. Mexico's remittances alone totaled over $25 billion in 2015, or around 2 percent of its total GDP. Most remittances in Middle and South America originate in the United States.

As countries in this region have sought to increase development, they have faced several challenges. Exports continue to flow from Latin America and the Caribbean to the rest of the world, though often coming at the cost of economic diversification. The island nations of the Caribbean have had particular challenges to sustainable development due to their small size and populations and limited natural resource base. These countries are referred to as **Small Island Developing States**, or SIDS (see **Figure 5.16**). These countries have struggled with high technology, communication, energy, and transportation costs and have had difficulty developing in a way that doesn't harm their fragile ecosystems. In the Caribbean, the SIDS have formed the Caribbean Community, or CARICOM, aimed at promoting economic integration and cooperation among its member countries.

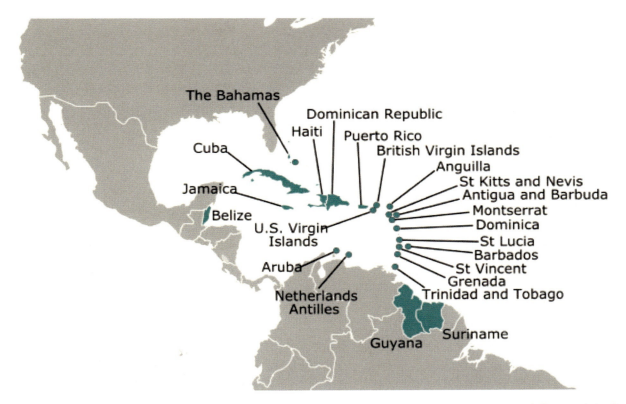

Figure 5.16: Map of the Small Island Developing States of Middle and South America (Derivative work from original by Osiris, Wikimedia Commons)

Some countries, particularly those in the Caribbean, have advanced their economies through **offshore banking**. Offshore banks are located outside a depositor's country of residence and offer increased privacy and little or no taxation. When the wealthy utilize offshore banks, they can thus avoid paying taxes on income that would be otherwise taxable in their home country. Belize, Panama, the Bahamas, the Virgin Islands, and many other countries in the region have become popular locations for offshore banks. The Cayman Islands, though, is one of the world's leading offshore banking locations. Around $1.5 trillion in wealth is held in the Cayman Islands and the British territory has branches for 40 of the 50 largest banks in the world. As a result, it has a GDP per capita of over $49,000 compared to just $8,800 for its much larger neighbor, Jamaica. Some countries have tried to strengthen their tax laws to prevent tax evasion through offshore banking.

Others in the region have turned to the production and trade of illicit drugs as a way to generate income, particularly cocaine and marijuana. Coca, the plant used to make cocaine, is grown and harvested in the Andes Mountain region, particularly in Bolivia, Columbia, and Peru. In 2013, Peru overtook Columbia as the global leader in cocaine production. The drug trade in Middle and South America has led to the rise of cartels, criminal drug trafficking organizations, and widespread violence in the region. Cartels often fight each other for territory, with civilians in the crossfire, and in many areas, drug organizations have infiltrated police, military, and government institutions. In Mexico alone, the ongoing Mexican Drug War between the government and drug traffickers shipping cocaine from Central America to global buyers has killed more than 100,000 people. The United States continues to be the largest market for illegal drugs. Americans purchase around $60 billion in illegal drugs annually, funding drug violence and drug trade in Middle and South America.

As countries throughout Middle and South America have increased their development, there have been some significant environmental concerns, particularly deforestation. When urban areas expand, forests are often cleared to make room for new housing and industry. Similarly, as agricultural lands expand and commercialize to feed growing populations and produce crops for export, it often leads to deforestation. Furthermore, nutrients in soil decline over time without careful land management, and thus after lands are intensively farmed for some time, soil fertility declines and new agricultural lands are cleared. Around 75 percent of Nicaragua's forests have been cut down and converted to pasture land. The Amazon rainforest, which amazingly holds around 10 percent of the entire world's known biodiversity, is down to around 80 percent of its size in 1970. The majority of the deforestation in the Amazon has occurred as a result of the growth of Brazil's cattle industry and its global export of beef and leather.

Despite slowing rates of deforestation and strides to address income inequality, this region remains largely in the global periphery. Some argue that it is to the advantage of countries like the United States to keep this region in the periphery, as it allows them to import cheap products. This idea is known as dependency theory, and it essentially states that resources flow from the periphery to the core, and thus globalization and inequality are linked in the current world system. While some have critiqued the specifics of the theory, others still see it as a useful way to understand the relationship between the core and the periphery. As Middle and South America continue to develop, they will face new challenges of how to do so in a way that is both ecologically and socially sustainable.

CHAPTER 6

Sub-Saharan Africa

> **Learning Objectives**
>
> - Identify the key geographic features of Sub-Saharan Africa
> - Describe the pre-colonial history of Sub-Saharan Africa
> - Explain the process of colonization in Sub-Saharan Africa and its effects on the modern geographic landscape
> - Analyze how colonization has impacted political stability and economic opportunity across Sub-Saharan Africa

6.1 THE PHYSICAL LANDSCAPE OF SUB-SAHARAN AFRICA

Africa is the cradle of human civilization. Our early ancestors, *homo erectus*, meaning "upright man," first walked in East Africa between one and two million years ago. Early humans in Africa were the first to create tools, develop language, and control fire. The physical landscape of Africa and its long history of habitation have contributed to a variety of cultures and human experiences.

Africa is the second-largest continent after Asia and is the only continent that is crossed by both the Tropic of Cancer, located 23 degrees north of the Equator, and the Tropic of Capricorn, located 23 degrees south of the Equator (see **Figure 6.1**). These tropics are areas of high atmospheric pressure creating dry conditions. The Sahara lies along the Tropic of Cancer in the north and the Namib Desert is situated on the Tropic of Capricorn in the south. The Sahara stretches across much of northern Africa creating a formidable barrier and dividing Africa between a Muslim, Arab North and traditional African cultural groups in the south. Since North Africa is so similar to Southwest Asia in terms of culture and political history, the two are discussed together in a separate chapter.

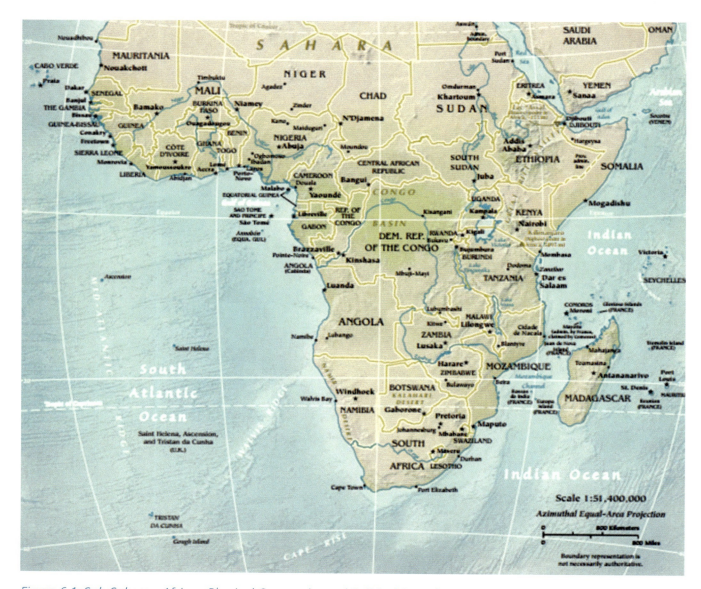

Figure 6.1: Sub-Saharan African Physical Geography and Political Boundaries (CIA World Factbook, Public Domain)

The story of Africa's physical geography begins 300 million years ago with the landmass known as Pangaea, the last supercontinent (see **Figure 6.2**). Around 175 million years ago, Pangaea began to break apart, drifting and colliding and forming the continents as we know them today. Africa was situated at the heart of this supercontinent.

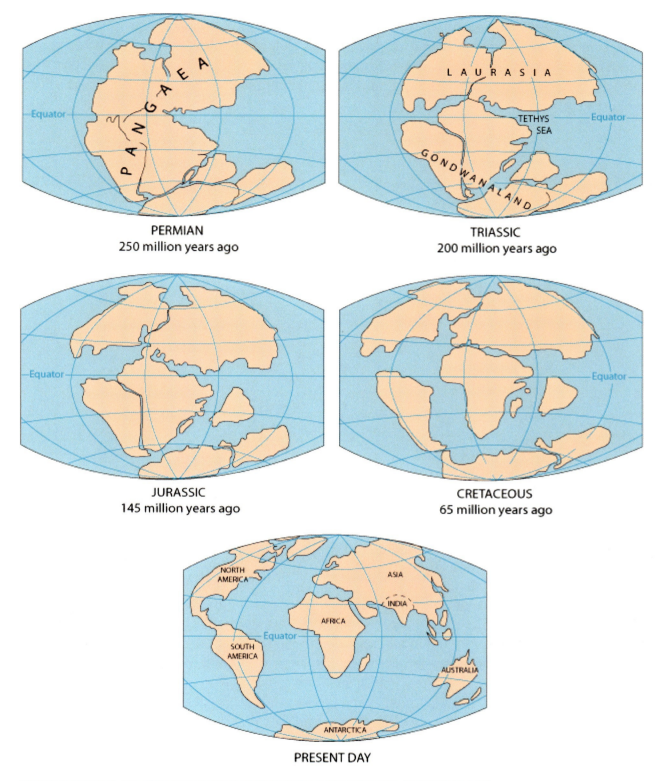

Figure 6.2: Break Up of Pangaea (United States Geological Survey, Public Domain)

Today, many of the physical landforms found in Africa were formed from this tectonic plate movement. Africa's Great Rift Valley, for example, is slowly splitting away from the rest of the African Plate at a rate of around 6 to 7 mm (around 0.25 in) each year (see **Figure 6.3**). That might not sound like a lot, but after 100 years, the rift would have expanded by two feet! Some of the deepest lakes in the world are found along this rift valley, where huge cracks in the earth's surface have filled with water over time. Lake Tanganyika, for example, is the second-largest and second-deepest freshwater lake in the world, dipping down to 1,470 m (4,820 ft). East of the rift valley is the **Horn of Africa**, a protruding peninsula that contains the countries of Djibouti, Eritrea, Ethiopia, and Somalia.

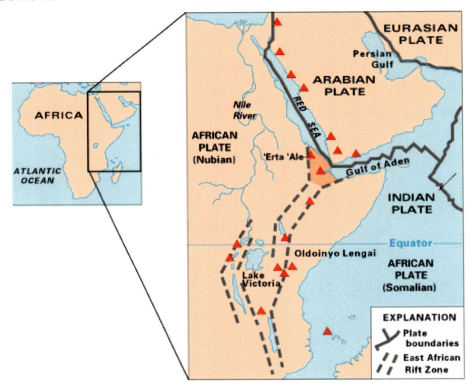

Figure 6.3: Map of Africa's Great Rift Valley (United States Geological Survey, Public Domain)

In addition to the rift valley, Sub-Saharan Africa contains a number of highland and plateau regions as well as large, tropical basins, the largest of which is the Congo Basin. This basin begins in the highlands of the rift valley and is the drainage area for the Congo River, Africa's largest river by discharge and the deepest river in the world. This watershed is considered a biodiversity hotspot and its forests support around 40 million people. However, there is serious concern in the region regarding deforestation.

Africa's other major river, the Nile, flows from Lake Victoria in the rift valley north through 11 different countries. It is regarded by most as the longest river in the world. The Nile has, historically and in modern times, been a key way to transport people and goods throughout the region and its floodplain enables farming in an otherwise arid environment.

Perhaps the largest ecoregion of Sub-Saharan Africa is the **Sahel**, located just south of the Sahara (see **Figure 6.4**). The Sahel is a transitional region connecting the dry Sahara to the tropical regions of the south. It is mostly grassland and has traditionally supported semi-nomadic livestock herders.

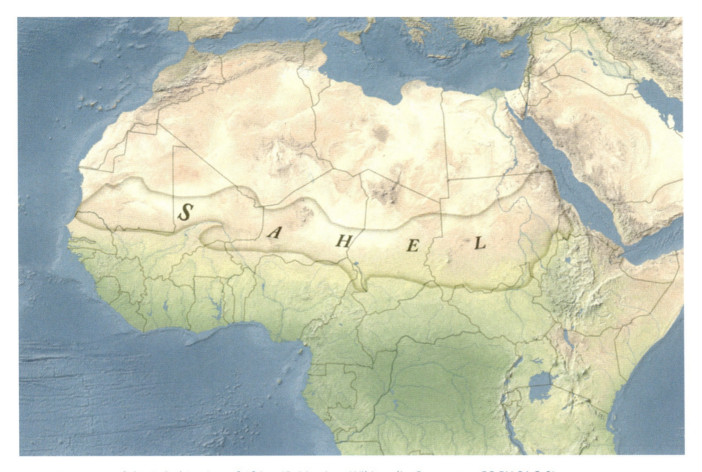

Figure 6.4: Map of the Sahel Region of Africa (© Munion, Wikimedia Commons, CC BY-SA 3.0)

The Sahel is at the front line of one of the most pressing environmental concerns in Africa: desertification (see **Figure 6.5**). **Desertification** refers to the process of previously fertile land becoming desert and occurs for a variety of reasons including climate change and human activities. Overgrazing, for example, can rid land of vegetation causing the erosion of fertile topsoil. Warming temperatures due to global changes in climate can change precipitation patterns and increase the speed of evaporation. Desertification in the Sahel has caused the Sahara to expand and has led to conflict as northern farmers have migrated to the south in search of fertile soil.

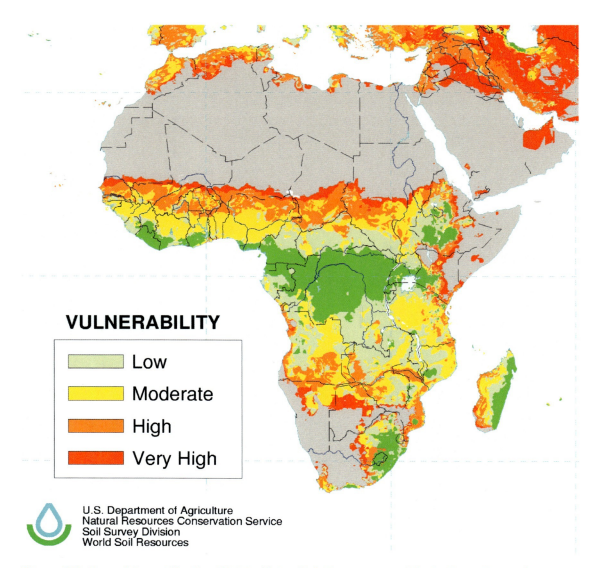

Figure 6.5: Map of Desertification Risk in Africa (U.S. Department of Agriculture, Natural Resources Conservation Service, Public Domain)

In addition to an array of landforms from rift valleys to mountains to deserts, Sub-Saharan Africa contains a wide variety of climate zones and precipitation patterns. In general, the continent is relatively hot with temperate climates in the higher elevations. Some areas of Sub-Saharan Africa, particularly the tropical rainforests of West Africa, receive upwards of 3,000 mm (118 inches) of rain each year, while other areas such as the Namib Desert receive less than 10 mm (0.39 inches) of rain annually.

6.2 PRE-COLONIAL SUB-SAHARAN AFRICA

Africa has a long history of human habitation giving rise to numerous cultural and linguistic groups. Early humans were primarily gatherers, and there is evidence of people gathering nuts, grasses, and tubers around 16,000 BCE in the highlands of Northern Ethiopia. Around 10,000 years ago, the domestication of the first crops and livestock developed in Africa and the practice of settled agriculture began.

In pre-colonial Africa, women were, and still are in many areas, the primary agriculturalists. It was the responsibility of women to understand the seasonality of crops and this, along with the role of bearing and rearing children, gave women an important role in African society. Many early religions placed a strong emphasis on female goddesses reflecting the central role of women in society. Men were primarily the hunters and gatherers.

For early Africans, the family was the basic and most important social unit. It was the family unit that owned and accessed land rather than individuals. Furthermore, land could not be bought or sold, but instead was passed down through the tradition of partible inheritance, meaning land is divided among the heirs. Elsewhere in the world, such as in the United Kingdom for much of its history, land was passed down to the firstborn male, known as primogeniture. With partible inheritance, however, no landed aristocracy developed since every male was given an equal share rather than just the firstborn.

If the family was the basic social unit in African society, it was the extended family that was the most important politically. **Tribes**, consisting of groups of families united by a common ancestry and language, controlled distinct tracts of territory. In pre-colonial Africa, there were over 800 distinct ethnic regions – and some of the ethnic regions identified by anthropologists actually had multiple distinct cultural groups within them. Tribal groups sometimes coexisted peacefully, and other times, warred over territory.

Pre-colonial Africa was also the site of a number of large empires (see **Figure 6.6**). The Kingdom of Kush, for example, was established in 1070 BCE and located on the Nile River just south of the Egyptian Empire. The Kingdom of Aksum located in present-day Eritrea and Ethiopia existed from 100 CE to 940 CE and was one of the most powerful of the pre-colonial African empires. The rulers of the kingdom minted their own currency, built religious monuments, and established trading routes. The first state in West Africa was the Empire of Ghana, which lasted from around 350 CE until its conquest by the Mali Empire in the 1200s CE. The Empire of Ghana had a large capital city, markets, and a system of taxation.

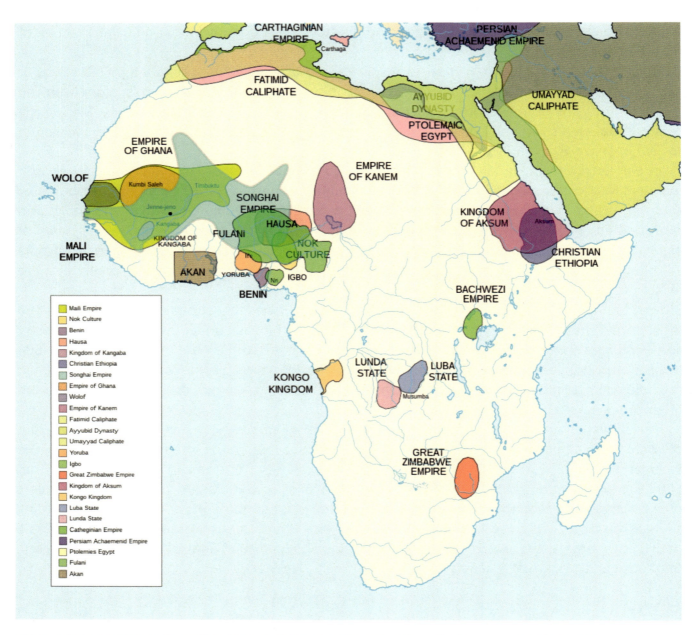

Figure 6.6: Map of Pre-Colonial Kingdoms in Africa (© Jeff Israel, Wikimedia Commons, CC BY-SA 3.0)

Two historical events brought significant change to Africa's cultural landscape and history. The first was the spread of the Islamic Empire across North Africa beginning in the 7th century CE. The second was the start of the transatlantic slave trade in the 15th century CE. Slavery was present in Africa long before European conquest, however. In some pre-colonial African societies, as in other parts of the world, slavery was a part of the local community. Slaves might be taken from conquered groups or given as gifts. In general, though, slavery represented a relatively small segment of ancient African society and economy.

European contact with Africa was initially focused on establishing a port along the West African coast, a place to resupply along the long trip to South and Southeast Asia. Beginning in the 15th century, however, this objective shifted to gaining resources. Portugal was the first of the Europeans to begin buying enslaved Africans. The Portuguese, originally interested in trading for gold and spices, set up several sugar plantations on the islands of São Tomé, off the western coast of Equatorial Africa. Portugal then brought in slaves to help cultivate the sugar. The Spanish then began buying slaves to send to the New World in the early 16th century, bringing them first to the islands of Cuba and Hispaniola. Initially, Europeans raided coastal African villages in order to secure slaves, but over time, began purchasing slaves from African rulers and traders.

Since pre-colonial Africa was divided into distinct ethnolinguistic regions, Africans generally saw themselves as part of their ethnic group or tribe rather than as united by a "black" or "African" identity. Thus, the slaves sold by Africans were seen as "other," something different or less than themselves. Europeans developed military alliances and beneficial trading partnerships with some of these groups to ensure a steady supply of slaves. By 1700 CE, around 50,000 slaves were being shipped out of Africa each year, and though it is difficult to know the exact total number of people sold into slavery, it is estimated that around 12 million Africans were shipped to the New World.

During this time, many African groups practiced a form of agriculture known as **shifting cultivation**, where one area of land is farmed for a period of time and then abandoned until its fertility naturally restores. Eventually, farmers return to the abandoned plot of land after many years, which is now overgrown, and often burn the vegetation, known as slash-and-burn, in order to return the nutrients to the soil. Because of this, much of the land in Africa looked unused, but was actually part of a larger agricultural system. Colonial empires took over these fragments of unused land to set up their own agricultural systems.

For some time, European empires had this relatively piecemeal approach to Africa, taking resources, land, and slaves without directly controlling territory. Europeans had little interest in the interior of Africa and were primarily focused on the coastal areas. This would completely change beginning in the late 19th century as European powers scrambled for control of the continent.

6.3 SUB-SAHARAN AFRICAN COLONIZATION

As the Industrial Revolution was spreading across Europe, colonial empires were seeking to expand their colonial holdings in order to gain mineral resources and expand agricultural production. As Europeans began exploring the interior of Africa, and recognized its resource potential, competition among European empires grew fierce. France, Italy, Britain, Portugal, and Belgium all raced through the interior of Africa trying to expand and strengthen their territories. When Germany entered the race, the colonial empires decided that it was in Europe's best interest to agree on and clearly demarcate African colonies and to agree on common policy.

In 1884, 13 European countries as well as the United States sent representatives to the Berlin Conference. At this conference, the colonial powers established the procedure for a Western country to formally control African territory and ultimately re-shaped the map of Africa. In a continent that had previously been divided into territories held by tribal groups and some larger kingdoms, the political boundaries were completely changed. By the early 20th century, around 90 percent of Africa was directly controlled by Europeans (see **Figure 6.7**).

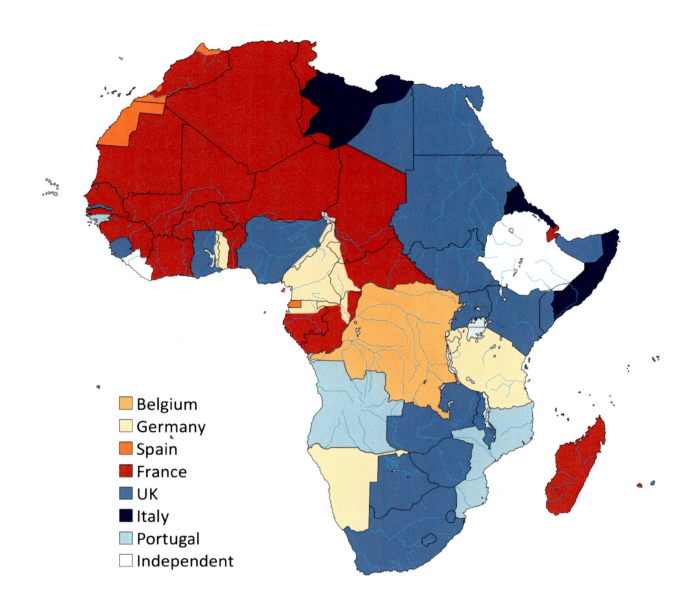

Figure 6.7: Map of Colonies in Africa, 1913 (Derivative work from original by Eric Gaba, Wikimedia Commons)

European colonization of Africa completely reshaped the political and ethnic landscape, with lasting effects even today. **Colonialism** broadly refers to the control of a territory by another group and colonial policies varied across Africa. In the Belgian Congo, what is now the Democratic Republic of the Congo, there was a racist ideology of paternalism where Africans were essentially viewed as children needing a fatherly, or paternal, authority to educate them in the ways of the West. In the far-reaching French colonies, from present-day Madagascar to Morocco, the colonizers emphasized an assimilationist policy, spreading the French culture through language, laws, and education. In the British colonies, like present-day Uganda, Ghana, and Nigeria, settlers partnered with local rulers who were made representatives of the British crown. This was known as indirect rule. The Portuguese, however, continued to be primarily interested in resources rather than local politics or culture. Their policy of exploitation in places like Angola and Mozambique ignored local development and the empire kept rigid control over local economies.

European colonizers were generally focused on exporting goods, with little attention given to local development or connectivity. These easily sold raw materials or agricultural goods are known as **commodities**. Local communities which may have previously practiced subsistence agriculture were shifted to export-oriented crops destined for European markets. When rail lines were built in Africa, they were generally constructed to simply take resources from the interior to the coastal ports without concern for developing regional linkages.

Eventually, the African colonies gained their independence, though the ease with which they were granted independence and the ease of their transition varied widely. Because the British practiced indirect rule, their colonies generally had a gradual transition of power with local rulers who'd been made representatives of Britain now governing over independent territories. Belgium, however, initially opposed the independence movement in its colonies, which were ruled directly by Belgian leaders. Most African countries would gain their independence following World War II. The last European colony to be granted independence was Djibouti in 1977.

While independence movements successfully broke territories free from European control, many areas were faced with a difficult decision regarding how to politically organize as a state. The European powers had redrawn the map of Africa with no regard for underlying ethnic territories. Some ethnic groups were thus grouped together, sharing colonial territory with groups they had long-standing conflicts with. Other ethnic groups were split apart, divided between two or more colonies. As the new political map of Africa took shape, it generally followed many of the colonial boundaries that had been artificially created.

Where two or more ethnic groups shared a newly independent territory, how should they decide who should rule? Perhaps the groups could simply share power and live peacefully. But what if one group, because of imposed colonial boundaries, no longer had access to resources? What if another group wanted to have unilateral control and not share it with groups they perceived as "other"? Many of the political and economic challenges facing the countries in contemporary Africa are rooted in its colonial history.

6.4 THE MODERN SUB-SAHARAN AFRICAN LANDSCAPE

Today, Sub-Saharan Africa is comprised of 48 independent countries and is home to 800 million people. While colonialism transformed African politics and economics, the way of life for many Africans has changed relatively little. Only around one-third of people in Sub-Saharan Africans live in cities, and as of 2007, 72 percent of these city-dwellers lived in slums. Sub-Saharan Africa is still largely rural. Urbanization is increasing, however, as governments have invested in industries in an effort to strengthen economic development drawing impoverished farmers from rural communities. Relatively few Sub-Saharan Africans live in large cities; most live in urban areas with fewer than 200,000 people. A notable exception to this is Nigeria, which was 48 percent urbanized in 2015 and contains several cities with over one million residents including its former capital Lagos (see **Figure 6.8**). The metropolitan area of Lagos has a population estimated to be 21 million; it is the most populous city in Africa.

Figure 6.8: Lagos, Nigeria (© Benji Robertson, Wikimedia Commons, CC BY-SA 1.0)

Sub-Saharan Africa's population has been climbing rapidly with the highest fertility rates of any region in the world. In Angola, for example, on the southwest African coast, most women have around six children. This has created a very high dependency ratio across the region, referring to the ratio of people not in the labor force to the number of productive workers. Africa's population is expected to double between 2010 and 2050. Nigeria, with 2017 population of over 190 million, is projected to surpass the population of the United States by 2050 to become the third-most populous country in the world.

In addition to rapid population growth, another significant challenge facing the governments of Sub-Saharan Africa is related to healthcare. Across the region, imbalances exist between the availability and quality of care. Furthermore, Sub-Saharan Africa's tropical climate has contributed to the spread of a number of serious illnesses, and this fact, combined with often ineffective or under-funded post-colonial governments have made stopping the spread of disease difficult. In some areas, western aid workers have been viewed with suspicion by Africans fearful of western intervention given their colonial histories. In other countries, foreign aid meant to help the poorest in the region has instead financed corrupt governments and military spending.

There are a number of illnesses, like hepatitis and hookworm, that are **endemic** to Sub-Saharan Africa, meaning they are found within a population in relatively steady numbers. Before a vaccine was developed, chicken pox was endemic in the United States, existing in a relative state of equilibrium. When a disease outbreak occurs, it is known as an **epidemic**. Epidemic diseases often affect large numbers of people on a regional scale. The flu in the United States is an example of an epidemic disease, with increasing numbers of people affected during the winter months.

Malaria, a disease spread by mosquitos, is the deadliest disease in Sub-Saharan Africa and sudden epidemics can affect large populations. If left without proper treatment, the disease can also reemerge months later. 90 percent of all deaths from malaria worldwide occur in Africa and it is estimated that the disease costs $12 billion each year due to increased healthcare cost, lost economic productivity, and a negative impact on tourism. Other insect-borne diseases that have a significant clustering in Africa include Yellow fever, which is also spread by mosquitoes, and trypanosomiasis, better known as sleeping sickness, which is transmitted by the bite of the tsetse fly.

When examining the geography of the global HIV/AIDS epidemic, Sub-Saharan Africa clearly stands out (see **Figure 6.9**). Around 70 percent of all people living with HIV/AIDS live in Sub-Saharan Africa and while a diagnosis with the disease might be met with long-term treatments in other regions, its fatality rate in Sub-Saharan Africa is much higher. In the hardest-hit countries, like Swaziland, Botswana, and Lesotho, more than one in five adults are infected. The disease is most often spread here by unsafe sexual practices, such as having sex unprotected with multiple partners even after marriage. Older adults in Africa are also deeply affected by HIV/AIDS, as they are often left to care for grandchildren orphaned by the disease. HIV/AIDS remains the leading cause of death in Africa.

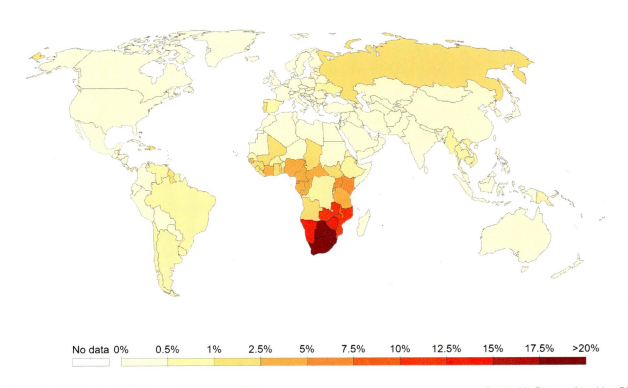

Figure 6.9: Map of the Share of the Population Infected with HIV by Country, 2017 (© Our World in Data, CC BY)

Still, there have been strides to try to slow the spread and assist those with the disease. Countries like Uganda have supported public awareness campaigns promoting monogamy in marriage and contraceptive use. They were able to slow the prevalence of HIV infections from 15 percent in the early 1990s to just 5 percent in 2001. In recent years, however, the infection rate in Uganda has once again climbed. Some say the government squandered aid money from the international community and did not keep up the public health initiative once rates decreased ultimately causing the infection rates to rise. There is also a concern that religious initiatives promoting abstinence have not had the desired effect. In a 2011 survey, around one-quarter of married men in Uganda reported having multiple sexual partners in the past year and just eight percent of them reported using a condom.

Periodically, regions in Africa, particularly West Africa, have experienced epidemics of Ebola, a viral hemorrhagic fever. Although Ebola is relatively difficult to transmit from person-to-person – it cannot be spread through the air like the flu – a combination of lack of understanding about disease transmission, inadequate infrastructure, and a distrust of Western intervention has made the region particularly vulnerable to deadly outbreaks. An outbreak beginning in 2013 in the coastal West African country of Guinea spread across the surrounding area and killed 11,000 people over the course of two years (see **Figure 6.10**).

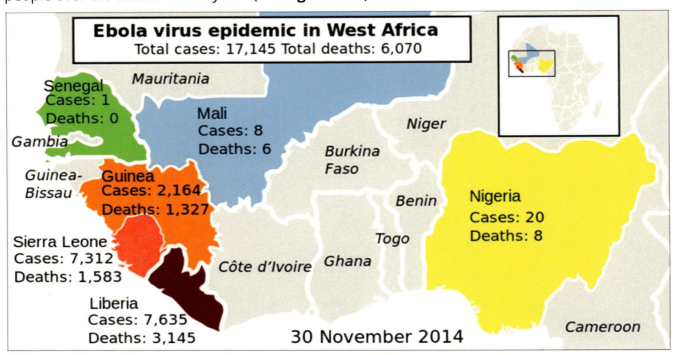

Figure 6.10: Map of the West African Ebola Outbreak, 2013-2016 (© Mikael Häggström, Wikimedia Commons, CC0 1.0)

The political issues and difficulties of dealing with Africa's healthcare crises illustrate larger issues related to governance in the region. Corruption is a significant problem across Africa costing residents around $150 billion each year. Bribery, even to access some public services, is common in many areas. Several governments deteriorated to the point where they are no longer functional, which is referred to as a **failed state**. In fact, of the ten states most vulnerable to failure in the world, seven are located in Africa.

In some areas, political conflict has gone hand-in-hand with ethnic conflict. Where competing ethnic groups found themselves sharing the same territory after independence, civil wars sometimes erupted as one group vied for power. Some groups turned to **genocide**, the systematic elimination of a group of people, in order to gain territorial and political control. In Rwanda, which had been a Belgian colony following Germany's defeat in World War I, the Belgian colonizers had given political power to the minority Tutsi. The Belgians perceived the Tutsi as closer to Caucasian, and this policy of granting power to minorities was a global colonial strategy aimed at suppressing the majority ethnic group.

After Rwanda gained independence in 1962 CE, the Hutu, who represented around 85 percent of the population, came to power and violent conflict between the Hutu and the Tutsi began almost immediately. Hundreds of thousands of Tutsi fled, becoming **refugees**, people who have been forced to leave their country. By the early 1990s, the Hutu began preparing for genocide, seeking to eliminate the Tutsi minority. Over the course of 100 days between April and July 1994, 800,000 Tutsis were killed – around 50 percent of the entire Tutsi population. By the time the United Nations admitted that "acts of genocide may have been committed," 500,000 had already been killed, most murdered with machetes. Rape was also systematically used against the Tutsi women. The Rwandan Patriotic Front, formed by exiled Tutsi refugees, defeated the government's forces and has governed Rwanda since the end of the genocide. Conflict between Tutsi and Hutu, particularly in neighboring Burundi, has continued, however.

In South Africa, the ruling party that came to power after independence was dominated not by an indigenous African group but by the descendants of Dutch settlers, known as Afrikaners. Africans in South Africa outnumbered non-Africans by 4 to 1 and the ruling Afrikaners instituted a policy of racial separation known as **apartheid** aimed at maintaining minority rule. This system of segregation led to an entirely different system of education for non-white South Africans, limited their use of public spaces, and forced the relocation of millions into separate neighborhoods. The system ended in 1994 with the country's first democratic elections, but inequality between white and non-white Africans persists, and has actually worsened according to some measures.

Some countries have tried to overcome political instability and the lack of colonial connectivity by creating interregional organizations aimed at economic and political cooperation. The largest of these organizations is the **African Union**, which formed in 2001 and consists of every African state, including Morocco which recently rejoined the AU in 2017. Among other objectives, the African Union seeks unity, integration, and sustainable development. Other regional organizations exist with narrower objectives, such as the Economic Community of West African States (ECOWAS) and the Common Market for Eastern and Southern Africa (COMESA), both of which aim to create free trade areas between member countries. Because of the diverse array of Africa's cultural and linguistic groups, organizations have often found it helpful to have a **lingua franca**, a common language spoken between speakers of different languages. In some cases, the lingua franca is the language of a common colonizer, such as English or French. In other cases, the language is native the region, such as with Swahili, the lingua franca for much of Southeast Africa.

6.5 ECONOMICS AND GLOBALIZATION IN SUB-SAHARAN AFRICA

Although formal colonization of Africa ended by 1980, in many areas, it was replaced with **neocolonialism**, the practice of exerting economic rather than direct political control over territory. During the colonial era, European groups formally controlled Africa's resources and created export-oriented economies. Today, most of Sub-Saharan Africa's exports remain raw materials. This makes the economies of countries in the region vulnerable to price fluctuations and global markets. Furthermore, although Western countries no longer directly control African land, corporations based in these countries have either bought land directly or invested heavily in the region. Investors have also purchased the water rights to some areas.

Neocolonialism is also evidenced on a broader scale. The peripheral regions of the world generally supply goods and labor used by the core countries. Thus, the core continues to exert economic pressure on the periphery to maintain beneficial trade partnerships and cheap products, labor, and raw materials. In many countries, particularly in Sub-Saharan Africa, a **dual economy** exists, where plantations or commercial agriculture is practiced alongside traditional agricultural methods. South Africa's dual economy, for example, has created dramatic differences in development within its own borders and has exacerbated income inequality in the country. Critics of neocolonial theory argue that too much blame is placed on colonialism for the modern economic problems in Sub-Saharan Africa. Indeed, government corruption, inefficiency, and internal exploitation remains a significant barrier to long-term economic development.

In some cases, economic pressure has come from loans granted by the core to the periphery. Although the intention of these loans was to assist the periphery in infrastructure development, many countries have struggled under the weight of staggering debt. Two lending organizations in particular, the International Monetary Fund (IMF) and World Bank, have loaned over $150 billion to countries in Africa. Loans by the IMF and World Bank to countries experiencing economic crises are accompanied by **structural adjustment programs** (SAPs), stipulating economic changes a country must make in order to make it better able to repay its loans. These conditions might include decreasing wages, raising food prices, or making the economy more market oriented.

In practice, though, structural adjustment programs limit the ability of states to make their own economic decisions, which some see as neocolonialism. In addition, SAPs require governments to drastically cut their spending which most often leads to cuts in social services and public health. These austerity measures can lead to economic stagnation, ultimately hampering a country's ability to pay back its loans – the very thing the SAPs were designed to do. Furthermore, aid packages and loan programs were generally designed to reflect Western ideas of development. This could be viewed as further evidence of neocolonialism in the region.

For some countries, the interest payments alone far exceed what they are able to pay. Globally, 39 countries, 33 of which are in Sub-Saharan Africa, have been identified as heavily indebted poor countries and eligible for debt relief through a joint venture by IMF and World Bank. Changes have been made to the program to specifically address poverty in these countries and shift the loan payment amounts to funding for social and public services.

Increasingly, African countries are partnering with investment groups rather than lending organizations and private investment now exceeds development assistance. China is Africa's largest trading partner and the country has invested billions in African infrastructure projects and direct aid. Chinese investment projects include a $12 billion coastal railway in Nigeria and a $7 billion "mini city" in South Africa. In Angola, Chinese investors built Kilamba New City, a town complete with 25,000 homes, schools, and commercial facilities, to be repaid by Angolan oil (see **Figure 6.11**).

Figure 6.11: Kilamba New City, Angola (© Santa Martha, Wikimedia Commons, CC BY-SA 3.0)

For low-income individuals in Sub-Saharan Africa, **microfinance** has provided a way to access a range of financial and investment services. These services might include small loans, known as microcredit. Women in particular have benefitted from microcredit, which does not require complex paperwork, extensive employment history, or collateral as with traditional loans, but has provided a way for women to become entrepreneurs. Interest rates on these loans is generally quite small, and over 95 percent of loans are repaid. Globally, microcredit loans totaled $38 billion in 2009. Globally, most of the 150 million microfinance clients are in Asia, but microfinance in Africa is growing.

Sub-Saharan Africa is one of the fastest developing regions in the world with a 4 percent growth rate in 2016 compared to a global average of 3.4 percent (see **Figure 6.12**). Much of Africa's economic growth has resulted from trade, and the region is rich in resources and has a large labor pool. However, despite significant mineral, agricultural, and energy resources, most people in the region remain impoverished. Africa's leading export, for example, is petroleum and petroleum products, and yet most Africans do not have access to a reliable source of electricity. Political instability, too, has limited the benefit of economic growth for many low-income Africans. Economic growth, if invested in infrastructure and public services, combined with increased educational and career training opportunities could improve incomes in the region. According to current estimates, by the end of this century, Africa's population will quadruple. This rapid population growth coupled with continued economic development could dramatically reshape the African geographic landscape.

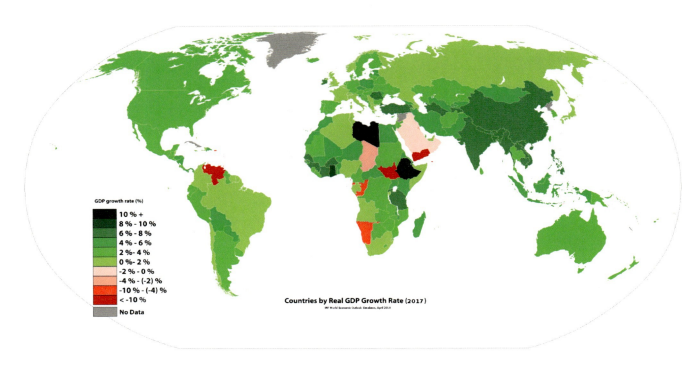

Figure 6.12: Map of Global GDP Growth Rates by Country, 2017 (© Kami888, CC BY-SA 4.0, Wikimedia Commons)

CHAPTER 7

North Africa and Southwest Asia

> Learning Objectives
>
> - Identify the key geographic features of North Africa and Southwest Asia
> - Describe the geography of the major religious groups found in North Africa and Southwest Asia
> - Explain how the history of North Africa and Southwest Asia impacted its cultural landscape
> - Describe the current areas of religious conflict within North Africa and Southwest Asia

NORTH AFRICA AND SOUTHWEST ASIA'S KEY GEOGRAPHIC FEATURES

When geographers divide the world into regions, we often do so using landmasses. Have a big chunk of land might be mostly surrounded by water? Let's make it a region! Sometimes, though, making these sorts of divisions is more difficult. Africa, for instance, is almost entirely surrounded by water except for a small land connection with Asia at Egypt's Sinai Peninsula. But Sub-Saharan Africa is physiographically, culturally, and linguistically distinct from the African countries north of the Sahara. In fact, North Africa has much more in common in terms of its physical and religious landscape with the Arabian Peninsula and Southwest Asia than some of its continental neighbors to the south (see **Figure 7.1**).

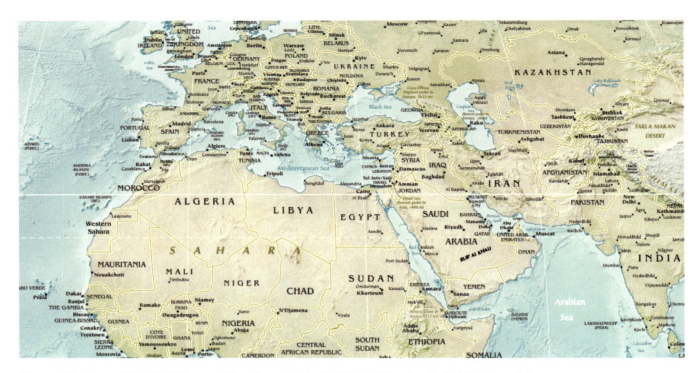

Figure 7.1: Map of North Africa and Southwest Asia (CIA World Factbook, Public Domain)

Historically, this perhaps awkwardly named region of North Africa and Southwest Asia was commonly called the "Middle East." This begs the question, though, what is it in the middle of? What is it east of? On a globe, east and west are relative terms. California is west of Europe but east of China. Indonesia is in Southeast Asia but is northwest of Australia. The equator might objectively be in the middle of the globe, but the "Middle East" is over 1,000 miles to its north. In truth, the term "Middle East" originated in Western Europe. Eastern Europe and Turkey were commonly referred to as the "Near East," while China was called the "Far East." The "Middle East" was thus in between these two regions.

Referring to the region as the "Middle East" seems to privilege the European perspective, so what alternatives exist? Perhaps you could call it the Islamic World? This would exclude places like Israel and secular governments like Turkey, as well as the numerous minority religious groups found in the region. You might have heard people refer to this area as the Arab World, but this would not apply to Iran, much of Israel, or Turkey. Thus we are left with simply the descriptive geographic name: North Africa and Southwest Asia, sometimes abbreviated as NASWA.

Whatever its name, this region is the hearth area for several of the world's great ancient civilizations and modern religions. The landscape of North Africa and Southwest Asia, as its naming difficulties imply, is marked by regional differences: in culture, in language, in religion, in resources, and in precipitation. Even within countries, regional
imbalances exist both in terms of the physical landscape and the patterns of human activity.

One of the most recognizable features of North Africa and Southwest Asia are its deserts. The Sahara, from the Arabic word ṣaḥrā' meaning "desert," is the largest hot desert in the world, stretching across 9.4 million square kilometers (3.6 million square miles) of the North African landscape. Although the typical image of the Sahara is its impressive sand dunes, most of the desert is actually rocky (see **Figure 7.2**).

Figure 7.2: Sahara, Algeria (© Cernavoda, Flickr, CC BY 2.0)

To the east, the Arabian Desert dominates the landscape of the Arabian Peninsula. In the southern portion of this desert is the Rub'al-Khali, the largest contiguous sand desert in the world. It is also one of the world's most oil-rich landscapes. There are also a number of highland areas across the region including the Atlas Mountains of Morocco, Algeria, and Tunisia and the Zagros Mountains of Iran, Iraq, and Turkey.

The prevailing climatic feature of North Africa and Southwest Asia is a lack of precipitation. From 10° to 30° north is a particular band of dry air that forms the region's hot desert climate zone (BWh in the Köppen climate classification system) and is clearly apparent on a map of global climate regions (see **Figure 7.3**). Most of the region receives less than 30 cm (12 in) of rain each year. This hot desert environment means that much of the land in the region is unsuitable for cultivation.

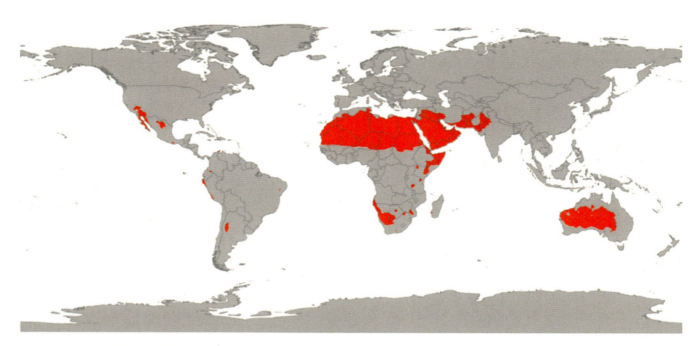

Figure 7.3: Map of Global Hot Desert (BWh) Climate Zones (© Peel, M. C., Finlayson, B. L., and McMahon, T. A., Wikimedia Commons, CC BY-SA 3.0)

There are exceptions to this arid environment, however. The region has a number of fertile river valleys and oases. The Nile River, for example, creates an arable floodplain in an otherwise extremely dry area (see **Figure 7.4**). While part of Iran is desert, northern Iran is actually home to dense rainforests and there are a number of scenic lakes. Coastal Turkey along the Mediterranean is often called the Turquoise Coast owing to its scenic blue waters. In general, however, those areas of North Africa and Southwest Asia where there is more abundant plant life are due to the presence of rivers, lakes, and seas rather than to the presence of ample rainfall.

Figure 7.4: Nile River Delta from Space (© Jacques Descloitres, MODIS Rapid Response Team, NASA/GSFC, Public Domain)

In a realm largely defined by its arid and hot climate, global changes in climate could have profound effects. Climate and physical geography have already significantly constrained human settlement and development patterns in North Africa and Southwest Asia. Rising temperatures could exacerbate droughts, and heat waves and dust storms will likely become more frequent. In some areas, conflicts over limited water resources have already begun. The Nile River, for example, runs through ten different states and 40 percent of the entire population of Africa lives within its floodplain. Egypt consumes 99 percent of the Nile's water supply, though, putting pressure on other countries, like Sudan, to keep water flowing downstream. Much of Egypt's water demand is for the irrigation of cotton, but cotton actually requires a significant amount of water and is a nontraditional crop for such an arid environment.

7.2 CULTURAL ADAPTATIONS IN NORTH AFRICA AND SOUTHWEST ASIA

The climate and physical geography of North Africa and Southwest Asia have shaped population patterns and culture in the region. People in the region are generally clustered around the region's sparse water resources reflecting ancient patterns of human settlement (see **Figure 7.5**). Four regions in particular stand out as having high population densities: the Nile River valley, the coastal Mediterranean Sea, the Euphrates and Tigris river basins, and valleys of northwestern Iran.

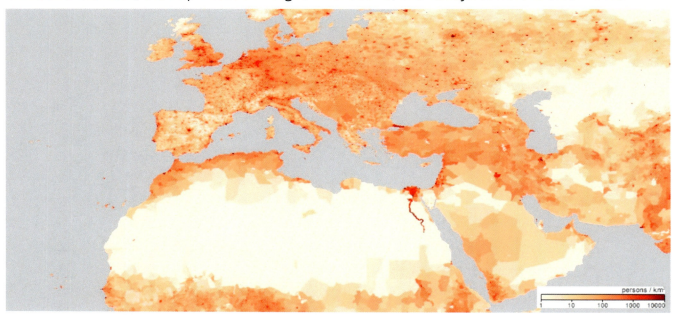

Figure 7.5: Map of Population Density in North Africa and Southwest Asia (© Robert Simmon, NASA's Earth Observatory, based on data provided by the Socioeconomic Data and Applications Center (SEDAC), Columbia University, Public Domain)

Over 10,000 years ago, the earliest humans in North Africa and Southwest Asia settled in the **Fertile Crescent**, the area surrounding the Tigris, Euphrates, and Nile rivers (see **Figure 7.6**). Here, humans first domesticated crops and animals and created the first farming settlements. In Mesopotamia, in particular, the river valley of the Tigris and Euphrates, innovations occurred that would change the trajectory of human existence. This was where the wheel was first invented, the first system of mathematics was created, and the first cereal crops, such as barley and wheat, were planted. Mesopotamia was also the site of the first urban civilization, called Sumer. Uruk, a city of Sumer, had a population of over 50,000 people by 2500 BCE making it the most populous city in the world at the time. The ancient city of Babylon, located between the Tigris and Euphrates, was inhabited for thousands of years and was likely the first city to reach a population of 200,000.

Figure 7.6: Map of the Fertile Crescent (© Nafsadh, Wikimedia Commons, CC BY-SA 4.0)

The people of this region have developed a number of adaptations to living in such a dry climate. Buildings are commonly designed with high roofs. Since hot air rises, having a higher ceiling allows the living area to remain relatively cool. Rooms are also often arranged around a common, shaded courtyard. This allows for maximum privacy, but also provides air flow throughout the living spaces. The traditional style of dress in parts of this region is also distinctive and reflects the physical landscape. Men might wear a cotton headdress to provide protection from the sand and sun as well as a long, flowing robe. Women's traditional clothing in the region is more reflective of religious values than environmental factors.

For some cultural groups in the region, adapting to the physical environment means migrating to cooler areas during the hottest parts of the year. The Berbers, for example, an indigenous group in North Africa, traditionally herd livestock and migrate seasonally seeking water, grazing land, and shelter. However, the way of life for many pastoral nomads in this region, like some Berbers, has changed significantly in recent years. Many governments have encouraged these groups to practice settled agriculture rather than seasonal migration, and international boundaries have often cut off traditional migratory paths.

The Persians, from modern-day Iran, devised an innovative way to transport water known as a **qanat**. Qanats are underground tunnels used to extract groundwater from below mountains and transport it downhill, where it is used to irrigate cropland (see **Figure 7.7**). They were developed over 2,500 years ago and many old qanats are still in use today in Iran as well as Afghanistan.

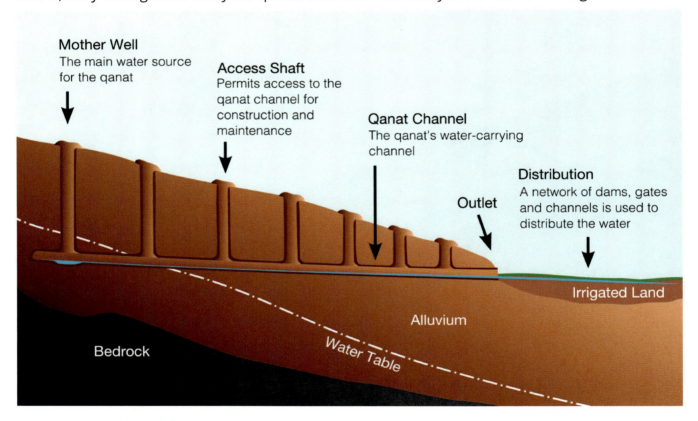

Figure 7.7: Cross-section of a Qanat (© Samuel Bailey, Wikimedia Commons, CC BY 3.0)

In such a harsh, arid environment, agricultural potential is fairly limited. River valleys and coastal areas provided small stretches of fertile land, but in the absence of widespread agricultural development, what other resources could bring this region wealth? In the early 20th century, oil was discovered in Saudi Arabia and this natural resource would prove both a blessing and a curse to the region.

Today, Saudi Arabia remains the world's leading oil exporter, shipping over 7.3 million barrels per day as of 2015. Kuwait, Iran, Iraq, and the United Arab Emirates are also among the top seven global oil exporters. Oil revenues have been able to increase development in these countries, financing industrialization, infrastructure, and providing high incomes. Qatar, for example, a small, former British protectorate on the coast of the Arabian Peninsula, has the highest GDP per capita in terms of purchasing power parity of any country in the world (as of 2018, according to the International Monetary Fund), at over $130,000 per person, largely due to its expansive oil and natural gas reserves. The tallest building in the world is now the Burj Khalifa, located in Dubai, United Arab Emirates. Although the United Arab Emirates, in building this 828 meter (2,717 feet) marvel, is seeking to diversify its economy and gain international recognition, its economy is still heavily dependent on oil.

Countries in the developing world with oil resources have often been prone to authoritarian rule, slow growth, corruption, and conflict. Oil wealth has been used to finance armies, and corrupt governments have pocketed oil revenue rather than reinvesting it in social programs or infrastructure. Furthermore, placing such a high emphasis on exporting one resource, like oil, has made this region vulnerable to changes in global energy demand. In 2015, countries in North Africa and Southwest Asia lost $390 billion in revenue due to low oil prices.

In an effort to coordinate oil production and prices, five countries including Venezuela, Iran, Iraq, Kuwait, and Saudi Arabia formed the Organization of the Petroleum Exporting Countries (OPEC) in 1960. Today, OPEC has 14 member states and covers over 40 percent of global oil exports (see **Figure 7.8**). OPEC cooperatively determines how much oil to produce and collectively bargains for the price of oil, rather than trying to compete to undercut one another. The United States and other countries have increased their own domestic oil production in recent years, causing OPEC's global share of oil exports to decline.

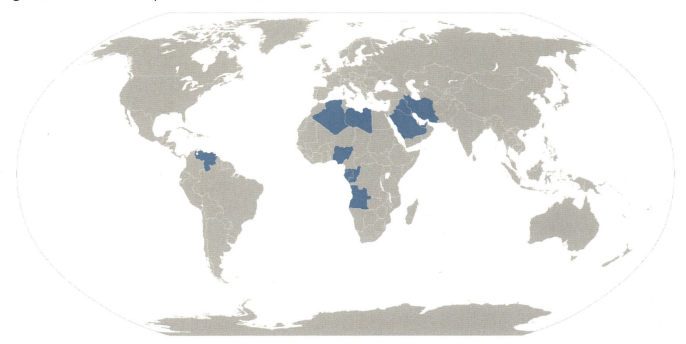

Figure 7.8: Map of OPEC Member States (Map by Bourgeois, Wikimedia Commons, Public Domain)

The presence of oil has also left a colonial legacy across North Africa and Southwest Asia, and has made these countries vulnerable to foreign control and influence. In addition, the uneven distribution of oil resources and wealth has led to inequality both within and between countries. Ethnic inequalities have also emerged as groups have uneven access to oil reserves and income.

Oil has also changed the pattern of human settlement in the region by bringing in migrants from outside the realm attracted by the prospect of economic opportunity.

7.3 THE RELIGIOUS HEARTHS OF NORTH AFRICA AND SOUTHWEST ASIA

North Africa and Southwest Asia is considered one of the great cradles of human civilization. It is also the hearth area for several of the world's major religions. These religions have changed the global cultural landscape, but have also led to tension and conflict throughout the region. Three religions in particular, Judaism, Christianity, and Islam, trace their ancestry through the tribal patriarch Abraham, who may have lived sometime in the 2nd millennium BCE.

The oldest of these Abrahamic faiths is **Judaism**. Judaism is a **monotheistic** religion, meaning that it is defined by a belief in one god. Jews believe that Abraham established the first covenant with God and the central Jewish text, the Torah, includes a discussion of the creation of the world as well as the establishment of this covenant. The first Jewish temple, built by Solomon, was constructed in modern-day Israel around 832 BCE. The Babylonians destroyed both the city of Jerusalem and Solomon's Temple in 587 BCE, and many scholars note that this event led to the creation of the written Hebrew Bible, what Christians term the Old Testament. The temple was reconstructed by Herod beginning in the 1st century BCE. However, this region has long been subject to invasion and conquest and in 70 CE, the Second Temple was destroyed by the Romans. This event prompted large-scale Jewish emigration from the region.

Both temples in Jerusalem were largely places of sacrifice and God was believed to literally dwell within the space. Thus, early Judaism was a temple-centered religion. The destruction of the Second Temple marked a distinct turning point in Jewish history between the historical Temple Judaism and modern Rabbinic Judaism. If a Jewish person is no longer defined by the sacrifices they make at the Temple, then what is it that makes someone Jewish? Rabbis and the interpretation of Jewish religious texts
became centrally important to creating new ideas of Jewish identity.

Today, there are around 14 million Jews worldwide; around 42 percent live in Israel, another 42 percent live in North America, and the rest live mostly in Europe. In addition, Judaism developed a number of different branches including Orthodox, which is more traditional, Reform, which is also known as Progressive Judaism, and Conservative, which is somewhere in the middle. Conservative is the largest branch of Judaism worldwide. However, millions of Jews around the world consider themselves to be unaffiliated or secular, emphasizing the ethnic and cultural values of the Jewish faith rather than the religious theology.

Christianity is another Abrahamic, monotheistic religion. It developed from the life and teachings of Jesus, a Jewish preacher who was born in 4 BCE in Judea, located in modern-day Israel. Jesus believed that the end times were near and emphasized love as the central religious doctrine. He was crucified by the Romans around 30 CE, a method of execution typically reserved for those who challenged the established social order.

Originally, Christianity was a sect of Judaism, but it eventually developed into its own, distinct religious tradition. A number of councils were held during the early years of
Christianity to create an agreed upon doctrine, though some of these decisions were disputed.

Over the years, Christianity developed distinct branches and denominations. The first of these divisions, known as the Great Schism, came in 1054 CE and was as much a product of geography as theology. This split divided the Eastern Orthodox from the Roman Catholic churches. In 1517 CE, the German monk Martin Luther penned *The Ninety-Five Theses*, which criticized Roman Catholic doctrine and began the Protestant Reformation.

Christianity is the largest religion in the world today with over 2.2 billion adherents. Although there are a wide variety of individual Christian beliefs, Christians generally view Jesus as a divine figure and believe he was resurrected following his death. Roman Catholicism remains the largest single denomination of Christianity with 1.2 billion members particularly in Brazil, North America, Western Europe, and parts of Africa and South America.

The religion that is most characteristic of North Africa and Southwest Asia today is **Islam**. Islam teaches in the existence of one God and emphasizes the belief in Muhammad as the last prophet. Followers of Islam are known as Muslims. Islam builds upon much of Jewish and Christian theology. Like Judaism, Islam views Abraham, Noah, Moses, and others as prophets of God. Also like Judaism, Islam has a monotheistic understanding of God; God is simply known in Islam as Allah, from the Arabic *al-ilāh*, meaning "the God." Islam teaches that Jesus was a prophet, and much of the story of his death and life in the Qur'an is similar to the story in the New Testament.

Muhammad was born in Mecca, in present-day Saudi Arabia, in 570 CE. Beginning when he was 40, Muslims believe that Muhammad began receiving revelations from God and later began preaching in his community. Muslims believe that the words of the **Qur'an**, the holiest book in Islam, contain the words of God as revealed to the Muhammad. Qur'an literally means "the recitation" in Arabic. In 622 CE, after widespread persecution, Muhammad was forced to emigrate to Yathrib, what is now the city of Medina, Saudi Arabia. This year marks the beginning of the Muslim calendar. In Yathrib, Muhammad gained converts and political authority and eventually the Muslim forces from Yathrib conquered Mecca, where Muhammad later died in 632 CE.

Immediately after Muhammad's death, disagreements arose over who should succeed Muhammad as the leader of the Muslim faith. Most Muslims believed that the leader
of Islam should be the person who is most qualified. Today, this group represents the Sunni branch of Islam. Others, however, believed that the only rightful leader must be a blood relative of Muhammad. This group is known as Shia, which derives from the Arabic phrase shi'atu Ali meaning "followers of Ali," who was Muhammad's cousin and son-in-law. Sunni is still the largest branch of Islam today, representing around 90 percent of adherents while Shia constitute around 10 percent of all Muslims (see **Figure 7.9**).

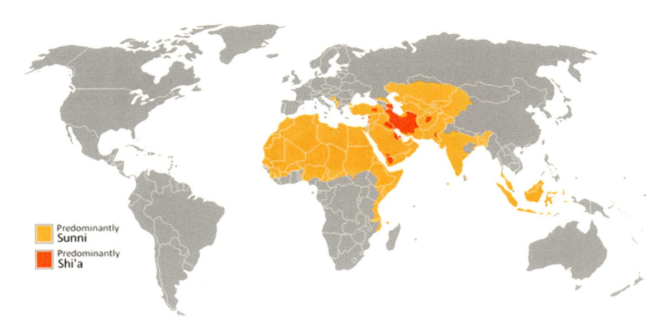

Figure 7.9: Map of Global Sunni and Shia Majorities (Derivative work from original by DinajGao, Public Domain)

The Five Pillars of Islam form the foundation for Muslim life and practice. First, the pillar of Shahada refers to a declaration of faith. It is generally recited in Arabic but translates as "There is no god but God (and) Muhammad is the messenger of God." Stating this phrase with conviction is all that is required to convert to Islam. The pillar of Salat refers to prayer five times per day. When Muslims pray, they face the Kaaba in Mecca, a cubed structure that is considered to be the most sacred Muslim site in the world (see **Figure 7.10**). The third pillar of Islam, Zakat, refers to the giving of alms, or charity. Muslims are required to donate 2.5 percent of all assets each year.

Figure 7.10: The Kaaba in Mecca, Saudi Arabia (© Muhammad Mahdi Karim, Wikimedia Commons, GFDL 1.2)

The fourth pillar, Sawm, requires Muslims to fast during the month of Ramadan. During this month, adult Muslims abstain from food, drink, and sex during daylight hours. Those who are ill, pregnant, elderly, or are otherwise unable to fast are exempt from the requirement. The month-long fast is designed to bring Muslims closer to God, but also to remind them of the feeling of hunger in the hopes that they will be mindful of those who are less fortunate throughout the rest of the year.

Finally, the fifth pillar of Islam is **hajj**, a pilgrimage to Mecca that is expected for all physically and financially able Muslims to complete at least once in their lifetime. During the hajj, which lasts several days, Muslims complete a series of rituals, some dating back to the time of Abraham. In 2012, a record 3.16 million pilgrims completed the hajj and crowd control has been a significant concern as the numbers of pilgrims have swelled. Since 1990, several stampedes have occurred; the deadliest was in 2015 and killed over 2,000 people.

Islam is the majority religion in every state in this realm except for Israel. Globally, Islam has around 1.8 billion followers and is the fastest-growing of the world's religions. Although this region is largely united by a belief in Islam, the divisions within the faith as well as the presence of numerous minority religious groups has often led to conflict.

7.4 CONQUEST IN NORTH AFRICA AND SOUTHWEST ASIA

After Muhammad's death, Arab military forces carried Islam across the region. At its greatest extent, the Islamic Empire under the Umayyad Caliphate of the 7th and 8th centuries stretched across 15 million square kilometers (5.79 million square miles), from the Iberian Peninsula, the southwest corner of Europe containing Spain and Portugal, all the way across North Africa and the Arabian Peninsula and into Pakistan (see **Figure 7.11**). No empire would be larger until the Mongols in the 13th century.

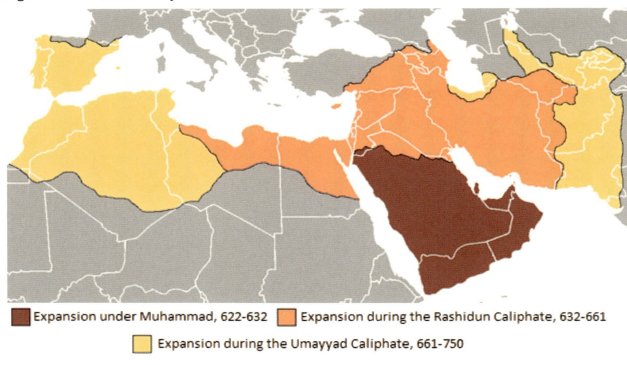

Figure 7.11: Map of the Islamic Empire under the Umayyad Caliphate Expansion, 622-750 CE (Derivative work from original by Brian Szymanski, Wikimedia Commons, Public Domain)

The Islamic Empire continued for hundreds of years. Its capital moved from Medina to Damascus, the capital of modern-day Syria, and then to Baghdad, the capital of modern-day Iraq. By 1259 CE, however, much of this region, including Baghdad, was conquered by the Mongols, beginning a pattern of occupation and conquest that would
continue until modern times.

The Ottoman Empire, based in modern-day Turkey, followed, taking control of much of North Africa and coastal Southwest Asia by the 15th and 16th centuries. Though it declined over time, the Ottomans maintained control of much of the region until it, along with the Central Powers of Germany, Austria-Hungary, and Bulgaria, lost World
War I. Following World War I, the allied powers of Europe divided the former territory of the Ottoman Empire and carved out colonies.

The League of Nations, an intergovernmental organization which lasted from the end of World War I until the beginning of World War II, portioned the former Ottoman Empire and granted mandates for European powers to control parts of its territory. France, for example, was given a mandate for Syria. Britain was given a mandate to control Iraq as well as Palestine. The Italians, too, were able to take a piece of the Ottoman Empire, gaining control of Libya in the early 20th century.

As with many other parts of the world, the colonies of North Africa and Southwest Asia were formed with little attention to underlying ethnic tensions or resource issues. Some ethnic groups found themselves split amongst several different European colonies, while others were forced to share newly created territories with hostile groups. Even once colonies were able to get independence, these colonial-era issues would remain. Unevenly distributed oil wealth, which was not discovered in large quantities until after the withdraw of European powers, would further complicate political and economic stability in the region.

7.5 THE MODERN POLITICAL LANDSCAPE OF NORTH AFRICA AND SOUTHWEST ASIA

Today's political map of North Africa and Southwest Asia reflects superimposed boundaries and a legacy of colonialization. The countries of this region have often been prone to political instability and conflict, and religious tension both between Muslims in this region as well as with the region's many religious minorities has often led to violence.

One key issue is that the geography of this region has often restricted development and transit to fairly narrow channels. Conflict can often occur over the control of these choke points. A **choke point** is a narrow passage to another region, such as a canal, valley, or bridge. North Africa and the Middle East has several, strategically important choke points including the Hormuz Strait, which provides the only sea passage into the Persian Gulf, and the Suez Canal, which was built to connect the Mediterranean Sea to the Red Sea. Who controls these choke points, and who they allow through, has often been a point of contention.

European colonizers were generally slow to relinquish control in the region. Local groups often reacted violently in trying to secure independence. As a result, many newly created governments in the region consisted of military groups. In other cases, monarchs found either military support or joined with local religious leaders. For many areas in this region, the discovery of oil brought about significant wealth, but also reignited Western interest and involvement. During the Cold War, for example, the United States sought to limit Soviet influence in the region and maintain its supply of oil.

Conservative religious ideology has sometimes provided a reaction against Westernization and foreign influence. In Iran, for example, the 1979 Islamic Revolution was largely a reaction against Westernization under a US-backed leader. The revolution established a **theocracy** in Iran, meaning a rule by religious authority, with the Grand Ayatollah, a Shia religious cleric, as the supreme leader.

After the dissolution of the Ottoman Empire, the Saud dynasty partnered with the leader of the Wahhabi religious movement, creating the foundation of modern Saudi Arabia. **Wahhabism** is a strict form of Sunni Islam that promotes ultraconservative Muslim values. Women have a strict dress code emphasizing modesty and have guardians, usually a father, brother, uncle, or husband, and need their guardian's consent to make major decisions or travel. Until 2018, women were forbidden from driving. A number of other practices are forbidden by Wahhabism, including watching nonreligious television programs, playing chess, and dancing. The penalties for breaking these prohibitions are often severe.

In Afghanistan, a group of militant Sunnis, known as **al-Qaeda**, fought against the Soviet Union's invasion of Afghanistan. The organization was founded by Osama bin Laden and formed an alliance with the Taliban, an Islamic fundamentalist political movement also based in Afghanistan. With al-Qaeda's military support, the Taliban were able to take control of Afghanistan from 1996 until 2001. They are known for their brutal oppression against women and acts of terrorism against civilian targets. As countries have modernized, Westernization and conservative religious values have continued to clash.

The landscape of North Africa and the Middle East remains in flux. The most widespread political change in recent years was a wave of protests and revolutions known as the **Arab Spring**. The Arab Spring began in Tunisia in 2010 when a fruit vendor set himself on fire after being harassed continually by police. Widespread protests in Tunisia followed his death, calling for changes to the country's issues with corruption, high unemployment, lack of political and personal freedom, and high food prices. After just ten days of demonstrations, Tunisia's president, Ben Ali, who had been in power for 23 years, fled in exile. From Tunisia, protests spread across the region, at times toppling governments that had been in power for decades (see **Figure 7.12**).

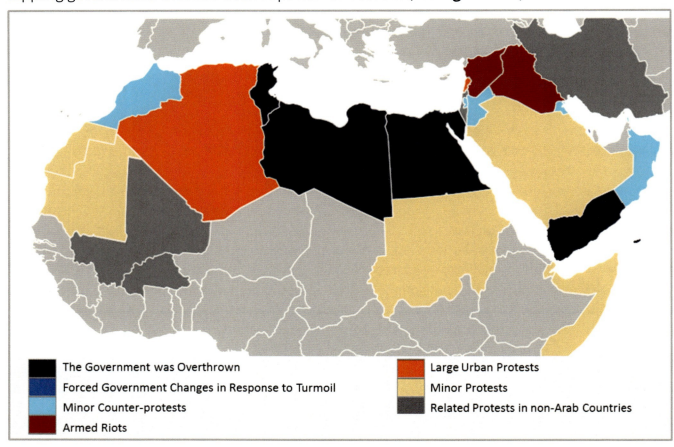

Figure 7.12: Map of the Arab Spring (Derivative work from original by Kudzu1, Wikimedia Commons)

At the heart of the causes behind the Arab Spring is inequality. In much of this region of the world, wealth and power is concentrated in the hands of a select few. Young people in the region in particular had high levels of education but also high unemployment and played a central role in bringing about change. Social media was also used to organize and rally support, and diffused the revolution rapidly. Several of the countries that experienced an initial change in regime have seen several later waves of political change, as interim governments sometimes proved to be as ineffective as previous leadership.

In Syria, however, despite widespread initial protests and calls for a change in leadership, president Bashar al-Assad not only refused to step down, but violently opposed protestors. Syria has been ruled by the Ba'ath political party, a socialist and nationalist group seeking Arab unity, since the 1960s. Bashar al-Assad was elected president under a 2000 referendum and ran unopposed, giving some indicator of the lack of political freedom in the country. Soldiers were ordered to open fire on civilian protestors and many were killed or tortured. Eventually, the country declined into civil war, with the government fighting rebel groups who sought to overthrow it and civilians caught
in the crossfire.

The civil war in Syria also offered an opportunity for another group in the region to gain control of territory. ISIS, the Islamic State in Iraq and Syria, also known as the Islamic State in Iraq and the Levant (ISIL) or just the Islamic State (IS), emerged in 2014 as a Sunni extremist group opposing the United States' invasion of Iraq. Iraq had been ruled by the minority Sunni population for centuries, but with the overthrow of then-president Saddam Hussein, also a member of the Ba'ath party, the majority Shia population took control. Efforts to create a coalition government and include Sunnis as well as the other minority groups in the country broke down. Some of the Sunnis who had been political leaders or military personnel under Saddam Hussein formed ISIS and were eventually able to drive out Iraqi government forces in several key cities.

From there, the group gained control of parts of Syria. For some time, much of Iraq and Syria existed as an **insurgent state**, a territory beyond the control of government forces. ISIS is widely known for its brutal tactics, including beheadings, sexual violence, and fundamentalist interpretation of Islam. The group sought to create a worldwide Islamic State with every Muslim country under its control. The United States declared ISIS defeated in 2019, though researchers note that while the US indeed took control of the last pieces of territories held by ISIS, thousands of ISIS fighters remain dispersed across Iraq and Syria and the group has the support of other affiliate groups and fighters across the world.

Since 2011, the people of Syria have endured government assaults, violence from rebel groups, and attacks from ISIS. Over 400,000 Syrians have been killed, many of them civilians, and over 13 million have become refugees. Some refugees have remained in Syria, cut off from aid by government and insurgent groups. Around 4.8 million people have left Syria. Some have fled to Turkey and Greece by boat; many have died on the perilous journey. Europe and North America have debated whether to accept these migrants, with some countries arguing that Syrian migrants might actually be terrorists, and others acknowledging that the global community has a responsibility to help those in need.

7.6 RELIGIOUS CONFLICT IN NORTH AFRICA AND SOUTHWEST ASIA

The rise of ISIS is representative of many key issues of geography in this region: the intersection of religious values, political instability, and control of territory and resources. ISIS represents a fundamentalist view of Islam, known as **Islamism**. Islamism is characterized by a strict, literal interpretation of the Qur'an, conservative moral values, and the desire to establish Islamic values across the entire world. Militant Islamist movements have inspired the violent ideology of **jihadism**, which seeks to combat threats to the Muslim community.

Islamism and jihadism represent a small portion of global Muslim beliefs, however. A Pew Research Center survey found that in most countries, over three-quarters of the population of Muslims reject Islamic extremism and a majority expressed concern over religious extremism. Furthermore, even in North Africa and Southwest Asia, only one-quarter of Muslims believed that tensions within the community between more religious and less religious Muslims represented a major problem. In every religious community, there are fundamental interpretations of scripture and both conservative and liberal understandings of theology.

For traditional Muslims, religious life and personal life are intertwined, and thus political structures in this region have often reflected religious values. Several states in North Africa and Southwest Asia have declared sharia law, meaning that Islamic religious law applies in the court system (see **Figure 7.13**). Islamist groups in particular have often utilized a strict and harsh interpretation of sharia. In many countries, most Muslims believe that sharia should be the state law, but many believe it should only apply to the country's Muslim population. In addition, Muslims differ in their interpretation of sharia, with some only supporting sharia for personal disputes but not accepting sharia's harsh punishments for various offenses.

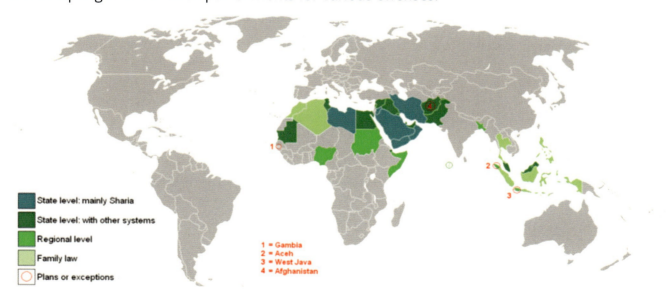

Figure 7.13: Map of Sharia Law by Country (© Sémhur, Wikimedia Commons, CC BY-SA 3.0)

In a region where political boundaries were often contrived by outside forces, governments have struggled with their relationship to minority religious and ethnic groups. The modern state of Israel, for example, was created following World War II by a United Nations plan to partition the British-controlled Palestine into Arab and Jewish areas. A series of wars between Israel and the surrounding Arab states left Israel as an independent state in control of the territories of the West Bank and the Gaza Strip, which were originally intended to be Palestine. Jerusalem, which under the UN plan would be a neutral, international city because of its significance to several of the world's religions, was proclaimed the capital of Israel.

Conflict continues between Palestinians living in territory controlled by Israel and Israelis who maintain sovereignty over the entire area (see **Figure 7.14**). Israel has built a series of walls dividing the West Bank and Gaza from the areas of Israeli control, maintaining that they are to protect Israelis from Palestinian terrorists. For Palestinians, however, these walls limit their freedom of movement and have often separated them from their livelihoods. The Gaza strip remains completely isolated, surrounded by walls on three sides and a sea controlled by Israeli ships on the other. Some have suggested a two-state solution and the creation of an independent, Palestinian state, but Israeli construction of homes in the West Bank has limited that option.

Figure 7.14: Map of Israel and the Palestinian Territories (Map by Scott, Wikimedia Commons, Public Domain)

North Africa and Southwest Asia is a region of the world that is the cradle of ancient civilizations and modern religions, but where resources are limited and unevenly distributed. Religious tension and political conflict have persisted. Some groups like ISIS have taken advantage of instability and valuable resources like oil to carve out control of territory and finance armed insurgencies. As some countries have modernized and industrialized, traditional religious values have often stood in stark contrast to the practices of migrant groups and tourists.

CHAPTER 8

South Asia

> **Learning Objectives**
>
> - Identify the key geographic features of South Asia
> - Explain the patterns of human settlement in South Asia
> - Describe the cultural landscape of South Asia
> - Analyze South Asia's current population growth and future prospects

8.1 SOUTH ASIA'S PHYSICAL LANDSCAPE

South Asia's Himalaya Mountains are the highest in the world, soaring to over 8,800 meters (29,000 feet). Yet, these are also some of the world's youngest mountains, reflecting a region that has experienced significant physical and cultural changes throughout its history. Here, we find one of the earliest and most widespread ancient civilizations, the hearth area for several of the world's great religions, and a region whose population will soon be the largest on Earth.

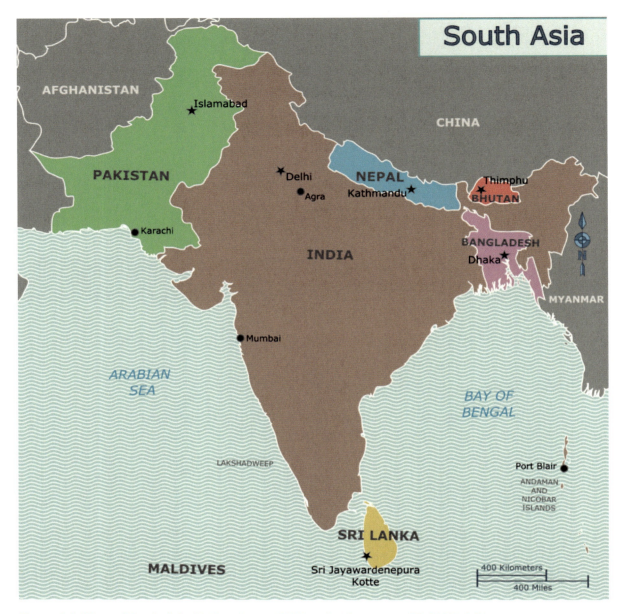

Figure 8.1: Map of South Asia (© Cacahuate, Wikimedia Commons, CC BY-SA 4.0)

South Asia is a well-defined region in terms of its physical landscape (see **Figure 8.1**). Formidable physical barriers separate the region from the rest of the Eurasian landmass. Much of the impressive physical geographic features of South Asia are the result of tectonic activity. Between 40 and 50 million years ago, the Indian Plate collided with the Eurasian plate (see **Figure 8.2**). Both the Indian Plate and the Eurasian plate were comprised of fairly low density material, and so when the collision occurred, the two landmasses folded like an accordion creating the mountain ranges we see today. The Indian Plate is still moving towards the Eurasian plate today and over the next 10 million years, will move an additional 1,500 km (932 mi) into Asia.

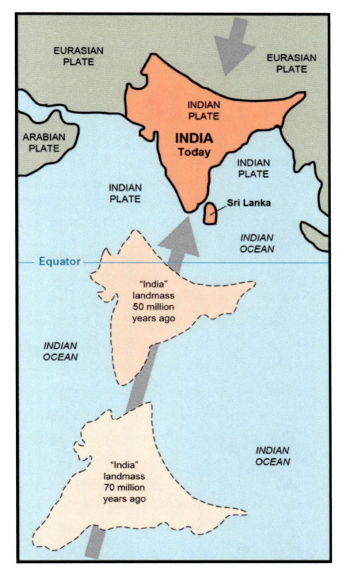

Figure 8.2: Indian Plate and Eurasian Plate Boundary (United States Geological Survey, Public Domain)

This massive tectonic collision resulted in perhaps the most well-known physical feature in South Asia: Mount Everest. Everest, located in the Himalaya Mountain range on the border of Nepal and China, is the highest mountain in the world. Because the India Plate continues to collide with the Eurasian Plate, this mountain range is still tectonically active and is rising at a rate of 5 mm each year. Thus, if you're planning on scaling Mount Everest in ten years, be prepared to climb an extra two inches.

Although the Himalaya Mountains are well-known for having the highest peak, the Karakoram Mountain range, passing through Pakistan, India, China, and Afghanistan, has the highest concentration of peaks above 8,000 meters (26,000 feet). Its highest peak, K2, is the second-highest mountain in the world and far fewer people have successfully made it to the top compared to Everest. One in four people die while attempting to summit.

Another key physical feature of South Asia, the Deccan Plateau, was also formed from the region's tectonic activity. Around 65 million years ago, there was an enormous fissure in Earth's crust which led to a massive eruption of lava. The entire Indian peninsula was buried in several thousand feet of basalt, a type of dense, volcanic rock.

South Asia's rivers, including the Indus, Ganges, and Brahmaputra form a lowland region that was home to several ancient civilizations. Today, these rivers provide for the water needs of many of this region's people, irrigation for agricultural lands, and an abundance of fish. However, these rivers have had significant environmental concerns in recent years and have supported increasing numbers of people along their banks. Most of the area along the Ganges River, for example, has been converted into urban or agricultural land and the wild species like elephants and tigers that used to be present along the river are now gone. Pollution in the Ganges River has reached unprecedented levels as industrial waste and sewage is dumped untreated into the river despite the fact that people frequently use the water for bathing, washing, and cooking. It is estimated that around 80 percent of all illnesses in India result from water-borne diseases. The World Bank has loaned India over $1 billion to clean up the river, but experts believe that larger scale infrastructure improvements are needed to improve the region's water quality.

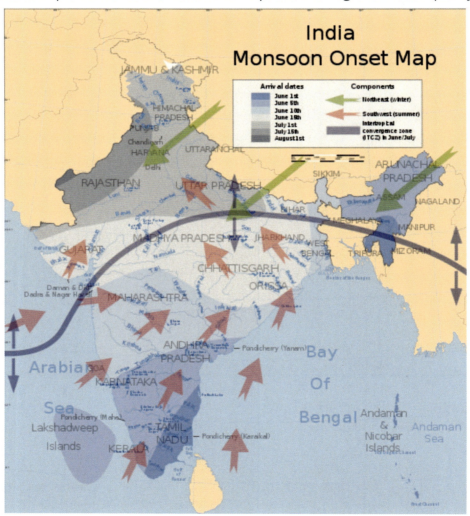

Figure 8.3: Map of Monsoon Onset Dates (© Saravask, based on work by Planemad and Nichalp, Wikimedia Commons, CC BY-SA 3.0)

The most important climatic feature of South Asia is a dramatic weather cycle known as the **monsoon** (see **Figure 8.3**). The monsoon refers to seasonal shifts in wind that result in changes in precipitation. From October to April, winds typically come from the northeast in South Asia creating dry conditions. Beginning in April, however, winds shift to the southwest, picking up moisture over the Arabian Sea, Indian Ocean, and Bay of Bengal.

Most of the rain during the monsoon season results from **orographic precipitation**, caused when physical barriers form air masses to climb where they then cool, condense, and form precipitation (see **Figure 8.4**). India's Western Ghats, a mountain range on its western coast, for example, causes orographic precipitation on its windward side. The Himalaya Mountains similarly result in orographic precipitation. However, these impressive highland areas are so formidable that they cause a dry area on their leeward side, known as a **rain shadow**. On one side of the Himalayas are some of the wettest places on Earth with over 30 feet of rain each year. On the other side, the rain shadow from the mountains forms the arid Gobi Desert and Tibetan Plateau.

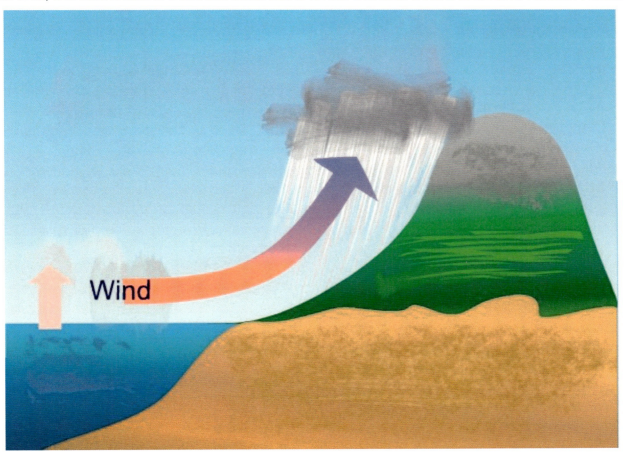

Figure 8.4: Orographic Precipitation (© Saperaud~commonswiki, Wikimedia Commons, CC BY-SA 3.0)

The monsoon rains, though extreme, provide significant benefits for South Asia's agriculture and economy. India gets more than 80 percent of its yearly rainfall from the monsoon and the rains are essential for both subsistence and commercial agriculture in the region. A good monsoon year will replenish the region's water supplies and increase crop yields, driving down food prices. Ample rainfall also contributes to the region's hydroelectricity potential. However, the torrential rains of the monsoon can also cause widespread flooding, destroying agricultural lands and transportation infrastructure, and can contribute to water-borne and insect-borne illnesses due to the significant amounts of standing water.

The monsoon is changing, though. Global changes in climate have made the monsoon harder to predict. In addition, rising numbers of automobiles across South Asia have increased air pollution, which can interfere with the mechanics of the monsoon. In the past, once the monsoon season starts, rains continue throughout the season. Recently, though, the monsoon rains have begun to stop and start throughout the rainy season. People in this region are generally unprepared for an unpredictable or variable monsoon season and rely heavily on the rains for agriculture. Local leaders are pushing for more research to better understand the shifting monsoon rains and for increased education on water conservation and sustainable agricultural management.

8.2 PATTERNS OF HUMAN SETTLEMENT IN SOUTH ASIA

South Asia's rich cultural landscape is a product of its varied physical environment and long history of human settlement. Modern humans first settled in this area 75,000 years ago, and early human ancestors likely settled in the region hundreds of thousands of years before that. The first major civilization in South Asia was in the Indus River valley beginning around 3300 BCE. This civilization, located in present-day Pakistan, Afghanistan, and northwestern India, relied on the monsoon rains to provide water to the Indus River. Here, early settlers developed systems of urban planning, baked brick houses, and the civilization at its peak numbered over five million people.

By 1800 BCE, however, the Indus Valley civilization began to decline. Weakened monsoon rains likely led to drought conditions and even small changes in precipitation and climate can have a devastating effect on a population of five million. Although residents developed some systems of water supply, they largely depended on the monsoon rains for agriculture, and many began moving to other areas of the region as arid conditions increased.

Around 1500 BCE, the Aryans, an Indo-Iranian group from modern-day Iran, invaded northern India. The Aryans were speakers of Indo-Iranian languages and brought their language, known as Sanskrit, their culture, and their ideas of social order to the South Asian realm. Hinduism and the caste system would both emerge from the Aryan culture.

South Asia was conquered by a number of different empires, each leaving an impact on the cultural landscape. The Maurya Empire stretched across the Himalaya and Karakoram mountain ranges, extending into most of South Asia by 250 BCE followed by a number of different dynasties. In the middle ages, the Islamic Empire extended into Afghanistan and Pakistan.

In the 18th century, however, the ruling Islamic Mughal Empire was in decline, leaving a power vacuum that would be exploited by the British. As the Industrial Revolution swept through the United Kingdom, the British were interested in expanding their supply of natural resources. Throughout the mid-18th century and the early 19th century, the British Empire, which had established the British East India Company, took over large stretches of land in India. The British established tea and cotton plantations, and took control of South Asia's resources. Although this region had previously established successful trading systems, the British saw local industries as competition and shifted their development to export raw materials. British rule also increased Westernization in South Asia and created an extensive rail transportation system.

As time went on, there were rising demands for independence. Mohandas K. Gandhi, known in India by the title "Mahatma," was a London-educated lawyer and one of the leaders in India's struggle for independence. He organized local communities to participate in nonviolent protests and his commitment to nonviolent resistance would inspire later civil rights leaders like Martin Luther King Jr.

Throughout this time, the isolated Himalayan countries of Nepal and Bhutan largely existed as **buffer states**, caught between the powerful British Empire and China. Their relative isolation allowed them to develop unique cultural features with little influence from outside groups, but as with most buffer states, left them with less economic and industrial development than their more powerful neighbors.

The British eventually agreed to withdraw from India but political and religious differences resulted in a **partition** of the former British territory in 1947 (see **Figure 8.5**). Areas that were majority Hindu would become the secular state of India. Areas that were majority Muslim would become the new Islamic state of Pakistan. Since Muslims were clustered both in modern-day Pakistan and along the mouth of the Ganges on the coastal Bay of Bengal, the Muslim state of Pakistan would be divided into a Western and an Eastern territory. This prompted large-scale migrations of Hindus and Muslims who were on the "wrong" side at the time of the partition.

Figure 8.5: Partition of British India and Migration (Derivative work from original by historicair, Wikimedia Commons, CC BY-SA 3.0)

Not everyone in South Asia supported the partition plan. Gandhi, who had long called for religious unity in the region, was opposed to the concept. Hundreds of thousands of people were killed in violent riots. In 1948, Gandhi was assassinated by a Hindu nationalist who opposed the partition plan and Gandhi's commitment to nonviolence.

Furthermore, although there were areas that were clearly majority Hindu or majority Muslim, religious minorities existed throughout India and not all regions had an easy transition. At the time of the partition, states were free to decide whether they wanted to join Hindu India or Muslim Pakistan. In the territory of Jammu and Kashmir in Northern India (see **Figure 8.6**), Muslims comprised around 75 percent of the population but the maharaja, the Sanskrit term for "great ruler," was Hindu. The maharaja struggled with the decision, and in the meantime, Muslim rebels, backed by Pakistan, invaded. He then gave the territory to India in exchange for military aid.

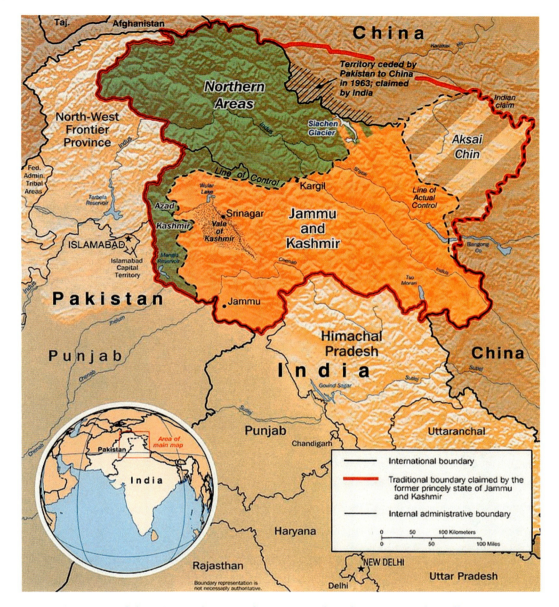

Figure 8.6: Map of the Disputed Areas of Jammu and Kashmir (Central Intelligence Agency, World FactBook, Public Domain)

Today, Jammu and Kashmir still remains a contentious territory and there have been violent clashes in the past few decades over political control. In the 1950s, China, without the knowledge of India, built a road through the northern portion of the state and was given territory by Pakistan. Although India claims the entire state, it controls the southern half of the state and about four-fifths of its population. Pakistan controls the territory's northern portion and moved its capital from Karachi to Islamabad to better control its frontiers. East Pakistan, long marginalized and culturally discriminated against by West Pakistan, gained independence as the state of Bangladesh in 1971.

As a region, South Asia is now the most populous area in the world and is home to over 1.8 billion people. Some of the world's largest megacities are located here as well, including Delhi, India (population of 26 million in the entire metropolitan area), Karachi, Pakistan (population of

14 million, with some estimating that it is much higher) and Mumbai, India (population of over 21 million). Despite the sizeable population, however, the region remains largely rural. Only around 36 percent of people in Pakistan, 31 percent of people in India, and 28 percent of people in Bangladesh live in cities. These relatively low levels of urbanization indicate that most people in the region still practice agriculture.

Urbanization is increasing, however, as industrialization and development have brought new jobs to the cities. British colonization left the region with the English language which has proven an economic asset, though it has also led to the marginalization of indigenous languages. Foreign companies have increasingly outsourced to India, taking advantage of a large, low-wage and English-speaking labor pool. **Outsourcing** refers to contracting out a portion of a business to another party, which might be located in a different country. Business processing in particular, such as call centers and information technology, has been outsourced and employs significant numbers of people in India. India is also one of the global leaders in fiber production, and textile production remains an important part of Pakistan's and Bangladesh's economies as well.

Nepal and Bhutan remain isolated both in terms of physical geography and global economic integration. Political uncertainty has generally hampered economic growth in Nepal but the country has been able to reduce its poverty rate considerably in recent decades. Tourism to Nepal has also increased, though local leaders have expressed concern over mounting issues of trash and pollution as a result of climbers flocking to Mount Everest. In the early 21st century, Bhutan transitioned from an absolute monarchy to a constitutional monarchy and held its first general election. Its government has promoted the measure of gross national happiness (GNH), as opposed to relying strictly on measures of economic or industrial development and has sought sustainable ways to develop and urbanize.

8.3 CULTURAL GROUPS IN SOUTH ASIA

South Asia is a diverse region in terms of its ethnic landscape, culture, and religious beliefs. As shown in **Figure 8.7**, in the northern portion of the region, the Indo-European languages like Hindi dominate as a result of the Aryan invasion. Along the Himalayas, languages in the Sino-Tibetan family dominate. In southern India, however, most groups speak a language in the Dravidian family, comprised of the indigenous languages of South Asia that were present before the arrival of the Aryans. These language families reflect broader differences in culture and ethnicity, including particular religious practices and food customs. Thus the label "Indian cuisine" actually encompasses a diverse array of regional and traditional specialties.

Figure 8.7: Map of the South Asian Language Regions (© Filpro, Wikimedia Commons, CC BY-SA 4.0)

South Asia is a hearth area for several of the world's great religions. Out of the Aryan invasion of northern India came a religious belief system known as Vedism. The religious texts of Vedism, known as the Vedas, combined with local religious beliefs developed into the modern-day religion of Hinduism by around 500 BCE. **Hinduism** is a polytheistic religion with a wide variety of individual beliefs and practices. Hinduism is a highly regional and individual religion and its polytheistic nature reflects this open understanding of belief. Of Hinduism's over 1 billion followers, 95 percent live in India.

At its heart, there are four key features of Hinduism: dharma, karma, reincarnation, and worship. Dharma refers to the laws and duties of being and is different for every person. You might be a student and an employee and a child and a sibling. All of those roles have prescribed responsibilities. To be a good student, for example, means to attend class, read the textbook, and study. In Hindu culture, there are also restraints and observances for how you interact with other people depending on their status.

Hindu views on the afterlife are quite different from the Judeo-Christian conception of heaven. Hindus believe in karma, which means that your deeds, good or bad, will return to you. They also believe in reincarnation, which is the idea that once you die, your spirit is reborn. Thus, you are the sum of numerous past existences. Karma, dharma, and reincarnation go hand in hand. If someone had done good deeds, had good intentions, and lived virtuously, when they die and are reincarnated, they might come back as something great – a prince, perhaps. Conversely, if someone was a terrible person, accumulating an excess of negative karma, when they are reincarnated, they might come back as someone of very low status – or maybe not even a person at all.

Hindu scripture discusses four distinct castes, or groups, of people in society, an example of **social stratification**. This social hierarchy is known as the **caste system**. The Brahmins, the highest caste, consist of priests and teachers and represent around 3 percent of India's total population. There is a warrior caste, a merchant caste, and finally the lowest, the laborer caste of landless serfs. Excluded from this caste system, and viewed as so below it that they are not even a part of it, are the "untouchables," also known as "Dalit" meaning "oppressed." The untouchables are so-named because they perform work that makes them spiritually unclean, such as handling corpses, tanning hides, or cleaning bathrooms. Traditionally, higher castes would get ritually purified if they touch a Dalit. Many untouchables are indigenous, non-Aryan Indians.

So how might the belief in karma and reincarnation affect social justice in South Asia? Although the caste system was outlawed by the Indian constitution, widespread discrimination and persecution persists. Many Hindus believe that those in lower castes were reborn into that social status because they had committed misdeeds in their past life. However, other Hindus fought against the caste system and have worked to more fully integrate the Dalits into Indian society.

Buddhism emerged out of Hinduism in northern India following the life and teachings of Hindu prince Siddhartha Gautama. According to Buddhist belief, Siddhartha lived a life of luxury, but became disenchanted with his life of privilege when he was faced with society's injustices, such as illness and extreme poverty. Since Hinduism offered no clear cessation of what Siddhartha viewed as an endless cycle of suffering through samsara, the soul's continual death and rebirth, he sought out new ways of ending suffering. For a time, Siddhartha practiced meditation and extreme asceticism, eating only dirt and bits of rice. But neither the path of luxury nor the complete absence of worldly pleasures gave him the insight he sought. Eventually, Siddhartha, in meditation under a Bodhi tree in Bodh Gaya, India, discovered what Buddhists refer to as the Middle Way, a path of moderation. He is said to have achieved enlightenment and is known as the first Buddha, meaning "awakened one."

Although Buddhism, like Hinduism, is a highly regional religion with many different forms of individual expression, Buddhists generally share a belief in the Four Noble Truths: 1) Suffering is universal and inevitable, 2) The immediate cause of suffering is desire and ignorance, 3) There is a way to dispel ignorance and relieve suffering, and 4) The eightfold path is the means to achieve liberation from suffering. Buddhists also share with Hindus a common belief in karma, dharma, and reincarnation.

Buddhism diffused across Asia, though never taking a strong hold in India (see **Figure 8.8**). The Maurya Emperor Ashoka, in particular, was responsible for the widespread diffusion of Buddhism in the 3rd century BCE. The religion has three primary branches, each with a distinct regional concentration. The oldest branch, Theravada, is primarily practiced in Southeast Asia, in places like Cambodia, Laos, Myanmar, and Thailand and is also the majority religion on the South Asian island of Sri Lanka. Mahayana is practiced by most Buddhists worldwide, particularly in places like China and Japan. Vajrayana Buddhism, which is sometimes considered a subset of Mahayana Buddhism, is practiced in the Himalayas and Tibetan Buddhism is a notable example. Buddhism has around 500 million followers worldwide.

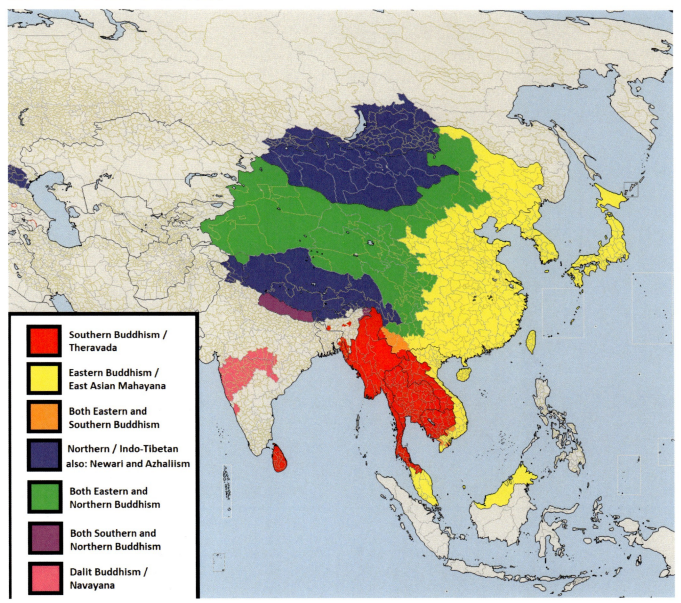

Figure 8.8: Map of Buddhism (© Javierfv1212, Wikimedia Commons, CC BY-SA 3.0)

Although Buddhism and Hinduism are the most widely practiced, South Asia was also a hearth area for the Jain and Sikh religions. **Jainism** emerged in India in the first century BCE and emphasizes ahimsa, nonviolence toward all living beings. Even insects found in the home are gently ushered out rather than killed. Jains also seek to break free from attachments and inner passions, and aim to keep an open mind toward different perspectives. The teachings of Jainism were influential for Gandhi and his emphasis on nonviolent resistance.

Sikhism emerged in the Punjab region of northwestern India and northern Pakistan in the 15th century. It is a monotheistic religion founded on the teachings of Guru Nanak that combines elements of both Hinduism and Islam. Like Hindus, Sikhs believe in reincarnation and karma. But unlike Hinduism, Sikhism prohibits the worship of idols, images, or icons. Sikhs believe God has 99 names, an adaptation of Hindu polytheistic belief. Sri Harmandir Sahib, commonly called the "Golden Temple," in Amritsar, India is the holiest Sikh temple, which are called gurdwara (see **Figure 8.9**). However, the building is open to everyone and every visitor is offered a free meal. Over 100,000 people visit the site every day.

Figure 8.9: Harmandir Sahib, Amritsar, Punjab, India (© Oleg Yunakov, Wikimedia Commons, CC BY-SA 3.0)

These religions, along with other minority religions like Christianity and indigenous belief systems, have not always coexisted peacefully in South Asia. Although India is officially secular, having no official religion, regional religious conflicts have often occurred throughout history. The difficulty is that in this region, very few people actually are secular, with no attachment to religion. Governments have thus struggled to find ways of accommodating minority religious groups while not offending the majority.

8.4 SOUTH ASIA'S POPULATION DYNAMICS

South Asia is the most populous region on Earth, but why is it the most populous, and how do geographers study population? The simplest way to measure population is to count the number of people in an area. India, for example, has a population of over 1.3 billion, making it the second-most populous country after China. But do raw numbers of people tell the whole story of the human population in an area? If two countries have the same population, but one is far smaller than the other, how could we examine population in a way that explores this difference?

Geographers often use the concept of density to investigate population. **Arithmetic density** is fairly easy to calculate. It is determined by simply taking the number of people in an area divided by the size of the area. If a territory was one kilometer square, for example, and was home to 100 people, the arithmetic density would be 100 people per square kilometer. Although arithmetic density is easy to calculate, it gives us a fairly limited view of population density. What if there are two tracts of land that are the same size and have the same number of people, but one is lush and fertile and has people spread out evenly and the other has a tiny river that everyone lives near? If you were using arithmetic density, the measurements for these two areas would be the same even though the actual settlement patterns are quite different (see **Figure 8.10**). **Physiologic density** takes into account this difference by examining the number of people per unit of arable, or farmable, land.

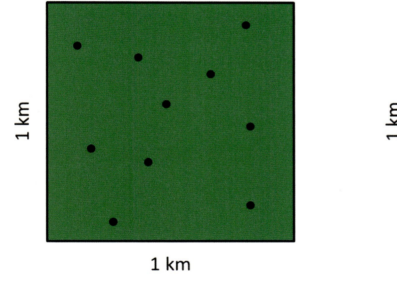

Figure 8.10: Arithmetic and Physiologic Density (Figure by author)

Arithmetic and physiologic density can give us insight into the concentration of a country's population and allows us to make comparisons between countries. The United States, a fairly large country, for example, has an arithmetic density of 32 people per square kilometer. However, a relatively small percentage of US land is arable, so the physiologic density is 179 people per square kilometer. Bhutan, by comparison, has a low population density of only 14 people per square kilometer. However, its rugged mountain environment means that only around 2 percent of the land is farmable, so its physiologic density is 606 people per square kilometer. By most measures, the most densely populated place in the world is Singapore with an arithmetic density of 6,483 people per square kilometer and a physiologic density of 441,000 people per square kilometer.

Another way to measure population is **agricultural density** which is the ratio of the number of farmers to the area of land. In developing countries where many people work as farmers, agricultural density is very high. South Asia has a high agricultural density. In developed countries, commercial agriculture and technological innovations have allowed relatively few people to be farmers and agricultural densities are generally low.

Geographers can also examine how a population is growing and changing over time. One way to explore this is with a **population pyramid**, a graphical representation of a population's age groups and composition of males and females. Ages of people are grouped in cohorts with younger people on the bottom and older on the top. Thus, a population pyramid that is very triangular has a lot of young people and is growing rapidly.

Typically, the ratio of males to females, known as the **sex ratio**, is 1 to 1 and population pyramids will have even sides. However, in populations where males are favored, the ratio may be skewed. Similarly, in countries where men have died in war, such as in World War II Germany, there might be more females. When geographers and population demographers refer to sex, it means something different from gender; sex is a person's biological identity as male or female while gender refers to a person's role as a "man" or "woman" within society.

India's 2017 population pyramid reveals rapid population growth over the past few decades (see **Figure 8.11**). However, the leveling off at the base of the pyramid indicates that population growth may be slowing. In addition, India's cultural preference for male children is clearly apparent. Among children aged zero to four, India has 62 million males and only 55 million females. Nationwide, there are over 47 million more males in India than females. Both abortion and infanticide have contributed to this imbalance.

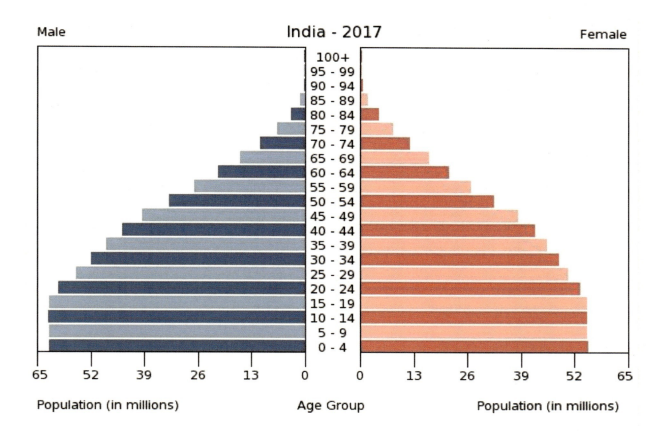

Figure 8.11: India's Population Pyramid, 2017 (U.S. Census Bureau, Public Domain)

All of the population pyramids for the countries in this region reveal preferences for male children, though none are as severe an imbalance as India. Although Pakistan's population growth has slowed in the past decades, its fertility rate remains the highest in the region at around 3.5, meaning a woman in Pakistan, on average, will have 3.5 children.

8.5 FUTURE CHALLENGES AND OPPORTUNITIES IN SOUTH ASIA

India's male-skewed population pyramid is indicative of a larger issue of gender inequality in its society. Sexual violence in particular continues to be a significant issue. Although the percentage of women who have been raped in India is lower than in other countries, a majority of rape cases are never reported and even an incidence rate of 8 or 9 percent in a population of over 1 billion people means that tens of millions of women have been victimized. The government of India has taken steps to reform its criminal code so that more criminals are prosecuted but even after a 2013 reform, marital rape continues not to be a crime. In a country that has few female police officers, high rates of domestic violence, and a relatively low status of women, sexual violence will likely remain a problem until these broader, systemic issues are addressed.

Overall, South Asia's growing population will have a significant impact on its geography. Much of the historic growth in this region was supported by the **Green Revolution**, which refers to changes in agricultural technology and productivity beginning in India in the 1960s. In 1961, India was at risk of widespread famine when a hybrid rice seed was developed that yielded ten times more rice than traditional seeds. It was called "Miracle Rice" and its use spread throughout Asia. Despite these agricultural advances, South Asia has the highest rates of child malnutrition of any world region. The low status of women in particular contributes to a lack of knowledge about the nutrients that are needed for children. Around one in three children in India are underweight.

Economically, South Asia has experienced rising prosperity yet systemic issues of governance and poverty remain. India in particular has one of the world's largest economies and the fastest growing economy in the region. This economic growth has mainly been focused on urban centers, drawing large numbers of people from the rural countryside to the cities in hope of finding work. Many cities have been unable to accommodate the rapid migration, however, and the sprawling slums in India, Pakistan, and Bangladesh are indicative of inadequate infrastructure and economic inequality. Several factories in this region have collapsed in recent years, killing thousands of workers and highlighting the poor working condition of many South Asians.

What does the future hold for South Asia? Although economic growth has reduced poverty in India, down from 60 percent in 1981 to 25 percent in 2011, corruption has increased. Inequality between genders, religious groups, castes, and ethnic groups remains a problem in much of the region. In some cases, this has led to **communal conflict**, which refers to violence between members of different communities. In Sri Lanka, a majority Buddhist country, ethnicity and religion are closely linked. Buddhists here have shaken the traditional peaceful image of their religion and have engaged in violent conflict with the minority Tamils and Muslims.

Still, local government and community leaders have sought to escape the shadow of the 20th century's turmoil by embracing new models of development and cooperation. In Bhutan, for example, the government initiative to measure gross national happiness resulted in shifting urban amenities, such as schools and healthcare clinics, to rural areas. This slowed the rural to urban migration that was rapidly occurring in other parts of the realm. Despite political and military turmoil, Pakistan has been able to substantially decrease its poverty rate. South Asia remains a complex realm at the crossroads of modernization and traditional cultural and religious values.

CHAPTER 9

East and Southeast Asia

Learning Objectives

- Identify the key geographic features of East and Southeast Asia
- Explain how East and Southeast Asia's history has affected its geographic landscape
- Describe the patterns of economic development in East and Southeast Asia
- Analyze how East and Southeast Asia interacts within the global economic system

9.1 THE PHYSICAL LANDSCAPE OF EAST AND SOUTHEAST ASIA

East and Southeast Asia (see **Figure 9.1**) contains the world's most populous country, the most populous metropolitan area, and some of the world's oldest civilizations. It is also a region with intense internal disparities and a landscape that has been and continues to be transformed by physical, political, and economic forces. Although East and Southeast Asia are often divided into two regions, they share a common economic and political history and global geopolitical forces continue to transform this realm.

Figure 9.1: Map of East and Southeast Asia (CIA World Factbook, Public Domain)

The region of East and Southeast Asia is divided from the rest of Asia by a number of formidable physical barriers (see **Figure 9.2**). In the north, Mongolia's Altay Mountains, the Mongolian Plateau, and the Gobi Desert separate the region from Russia. In the south, the Himalaya Mountains divide China from South Asia and contain the world's highest mountain, Mount Everest. These mountains are so high, in fact, that they form the Gobi Desert by preventing rainfall from passing over South Asia into Central Asia. In the southeast, the Arkan Mountains and Naga Hills, which stretch across Myanmar and India, and the rolling hills of China's Yunnan Plateau separate Southeast Asia from the rest of the continent. In general, this is a realm of relatively high relief, meaning there are significant changes in elevation on the landscape. Even the islands of this region have a rugged topography, from Japan's Mount Fuji to Indonesia's Mount Carstensz.

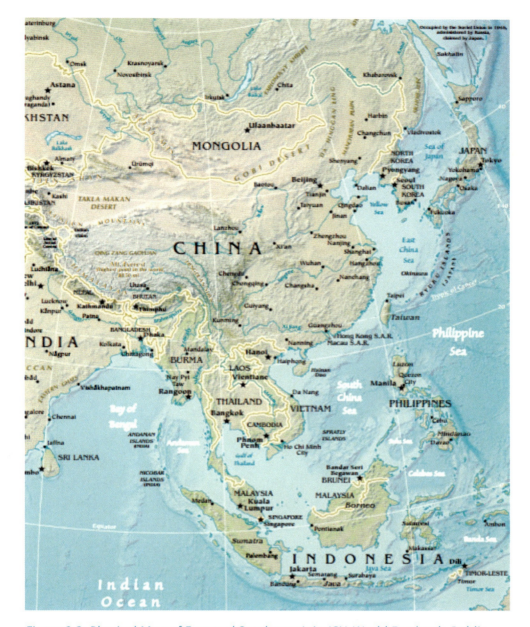

Figure 9.2: Physical Map of East and Southeast Asia (CIA World Factbook, Public Domain)

The rivers of this region have supported both ancient cultures and modern societies providing irrigation for agriculture, river transportation, and in some cases, hydropower. Asia's longest river, the Yangtze, flows through central China; the economic activity surrounding its river valley generates around one-fifth of the entire country's gross domestic product (GDP). In 2003, the Chinese government built the Three Gorges Dam, the world's largest hydroelectric power station, which spans the river. China's other major river, the Huang He River, also known as the Yellow River, flows through the highlands of Western China before discharging in Northeastern China. It was on the banks of the Huang He that Chinese civilization first began. In Southeast Asia, the region is dominated by the Mekong and Irrawaddy Rivers. The Mekong River, one of the most biodiverse rivers in the world, has been heavily dammed, impacting the area's ecology, and plans are underway to dam the Irrawaddy in several places. In addition, both the Mekong and the Irrawaddy originate in China, presenting issues over river flow and ownership.

Although the construction of the region's numerous dams has provided reliable power, they've been met with significant social and ecological impacts. The Three Gorges Dam, for example, was an unprecedented engineering marvel and will reduce the potential for downstream flooding, but flooding from the creation of the dam displaced over one million people and significantly reduced forest area around the river.

Most of the region's people live in the more temperate climate zones. In East Asia, for example, the coastal regions of Central and Southern China, Japan, and South Korea are primarily a humid temperate climate. Southeast Asia is largely tropical with ample rainfall throughout the year. The exception to these relatively warm areas are Western China, where the cold highland climate dominates, and Northeastern Asia is quite cold due to its high northern latitude.

The region's physical landscape has significantly affected its agricultural practices. The banks of East and Southeast Asia's rivers provided early settlers with fertile soil, and even today, provide agricultural irrigation. The region's hilly terrain, though initially an obstacle to agricultural productivity, inspired innovations such as terracing, cutting a series of flat surfaces resembling steps on hillsides. China in particular continues to be a global leader in terms of agricultural production.

9.2 NATURAL HAZARDS IN EAST AND SOUTHEAST ASIA

Much of what unites this region is its instability, not necessarily in terms of geopolitics, but rather its physical landscape. East and Southeast Asia are located in the **Pacific Ring of Fire**, an area of high tectonic activity along the Pacific Ocean basin (see **Figure 9.3**). The vast majority of the world's earthquakes, around 90 percent, occur along this geologically unstable area.

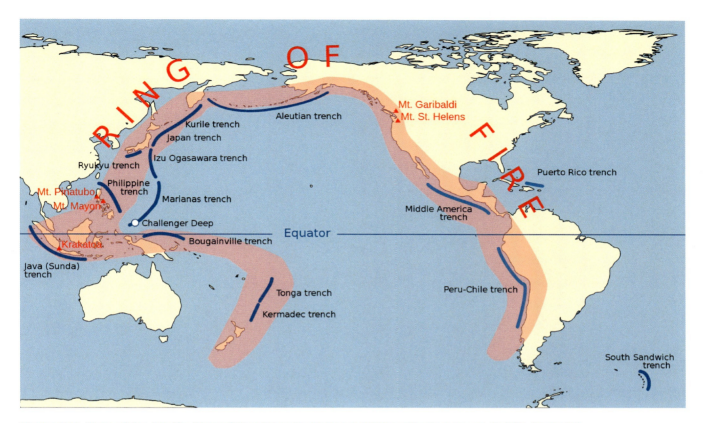

Figure 9.3: Map of the Pacific Ring of Fire (Map by Gringer, Wikimedia Commons, Public Domain)

In East and Southeast Asia, tectonic collisions have shaped the physical landforms present in the region and present numerous natural hazards (see **Figure 9.4**). Volcanoes erupt in this region frequently, and many of the islands in this region were actually formed from a variety of historic volcanic eruptions. Japan's highest mountain peak, for example, the majestic Mount Fuji, is an active volcano that last erupted in the early 18th century. The 1815 eruption of Indonesia's Mount Tambora, one of its dozens of active volcanoes, was so powerful, it cooled global temperatures and caused crop failures as far away as Egypt and France. In 1883, the volcanic island of Krakatoa, between the Indonesian islands of Java and Sumatra, erupted with such a violent explosion that it actually collapsed. Tens of thousands died and it took several years for global weather patterns to return to normal. The sound of the eruption is considered the loudest sound in modern history and could be heard from over 4,800 kilometers (3,000 miles) away.

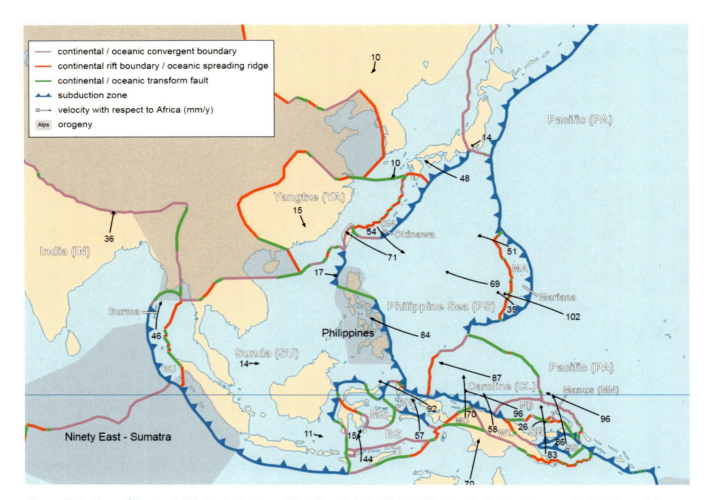

Figure 9.4: Map of Tectonic Plates in East and Southeast Asia (United States Geological Survey, Public Domain)

Along the islands of Indonesia, the Australian Plate is subducting, or moving below, the Eurasian Plate. This has resulted in a **subduction zone** west of the Indonesian island of Sumatra, an area of high seismic activity. In 2004, this subduction resulted in a massive undersea earthquake, so powerful that it actually shortened the day by a fraction of a second. The earthquake triggered a series of **tsunamis**, high sea waves, which devastated coastal communities in 14 different countries and killed 230,000 people. Tsunamis result from the displacement of water and can have a variety of causes, such as a landslide, meteor impact, or undersea volcanic eruption. Most commonly, though, they result from earthquakes.

In 2011, the most powerful earthquake to ever hit Japan, a magnitude 9.0, resulted in tsunami waves over 40 meters (131 feet) high in some areas. The earthquake and subsequent tsunami left over 15,000 dead and damaged hundreds of thousands of buildings. Most notably, the tsunami damaged Japan's Fukushima Daiichi Nuclear Power Plant causing a series of nuclear meltdowns and the release of radioactive material.

So what can be done about the danger from earthquakes, volcanoes, and tsunamis in this region? While nothing can stop Earth's massive tectonic plates from moving, warning systems, land use planning, and public education could help prevent casualties. After the 2004 Indonesian earthquake and tsunami, the international community created an Indian Ocean tsunami warning system. Although some of the hardest hit areas would have had only minutes to find higher ground, the warning system could have had a significant impact in alerting areas ahead of the wave. In Japan, earthquake drills are common and strict building codes ensure that buildings can withstand most seismic activity.

In addition to these geologic hazards, typhoons are also common in this region. Typhoons, the term for tropical cyclones in the northwestern region of the Pacific Ocean, routinely make landfall in East Asia. The region actually has more tropical cyclone activity than anywhere else on Earth. Most storms form in the summer between June and November and the islands of the Philippines are generally the hardest hit. Monitoring systems have been in place in the region for several decades and have helped to minimize the impacts from these powerful storm systems.

9.3 EAST AND SOUTHEAST ASIA'S HISTORY AND SETTLEMENT

The history of human settlement in East and Southeast Asia begins in China. Evidence of modern humans can be found in the region dating back to over 80,000 years ago. Around 10,000 years ago, several cultural groups emerged in China during the **Neolithic Period**, also known as the New Stone Age. This was a time of key developments in early human technology, such as farming, the domestication of plants and animals, and the use of pottery. Along China's Yangtze River, humans first domesticated rice around 6500 BCE. Villages, walled cities, and great dynasties, or families of rulers, emerged later.

While some early humans stayed in East Asia, others followed the coastline and continued on to Southeast Asia likely over 50,000 years ago. This was during the glacial period known as the Ice Age. Global temperatures were much colder and huge sheets of ice covered North America, Europe, and Asia. Since so much water was trapped
in these huge glaciers, ocean levels were actually much lower than they are today. Indonesia, Malaysia, and the other islands of Southeast Asia were a single landmass known as Sunda (see **Figure 9.5**). Those cultural groups who had seafaring knowledge continued on, populating Australia and the surrounding islands. During the Ice Age, the southern islands of Japan were also connected to the rest of Eurasia, allowing the indigenous groups of Japan to migrate from what is now mainland China.

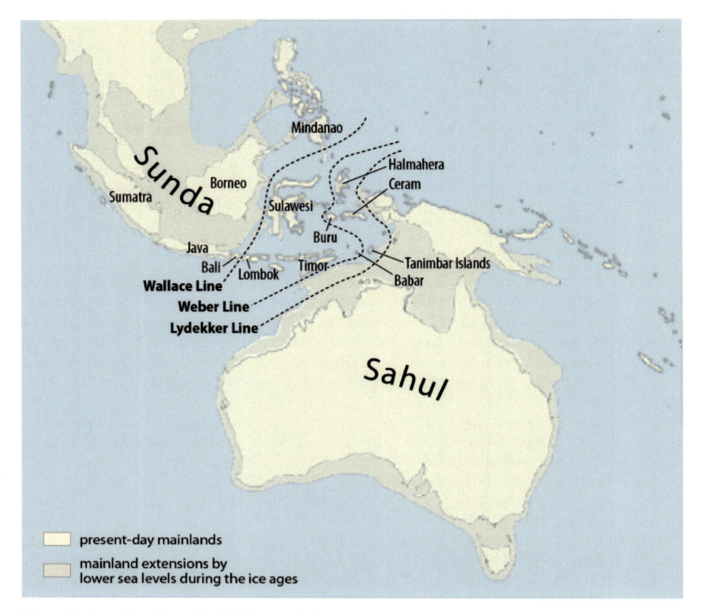

Figure 9.5: Map of Southeast Asia 20,000 Years before Present (© Maximilian Dörrbecker, Wikimedia Commons, CC BY-SA 3.0)

In East Asia, the Chinese dynasties dominated the political landscape for much of the region's history. They established trade routes, a strong military, and forged connections with Korea and Japan. China became a unified state under the Han dynasty, which ruled from 206 BCE to 220 CE, and this long period of stability is viewed as
a golden age in Chinese history. The dominant ethnic group in China, the Han, take their name from this ruling family. It was also during this time that Confucianism became the state religion. Confucianism takes its name from the influential Chinese philosopher and teacher Kong Fuzi (551-479 BCE), often referred to by the Latinized version of his name, Confucius. One of Confucius' key teachings was the importance of relationships, both within the family and within society as a whole and the religion emphasizes human goodness and self-reflection rather than the worship of

a divine being. Confucius also emphasized education and his teachings have dominated Chinese culture for centuries.

In general, the Chinese dynasties were largely isolationist. China has a number of physical barriers that separate it from the rest of Asia, such as the Himalayas, the rugged western highlands, and the Gobi Desert. The only region where it was vulnerable to invasion was its northeastern region. Here, the ruling families of China built a series of walls, known today as simply the Great Wall of China (see **Figure 9.6**).

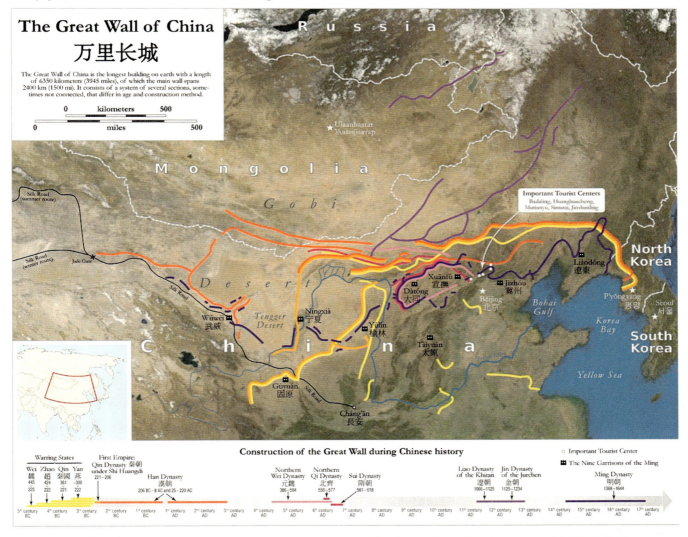

Figure 9.6: Map of the Great Wall of China (© Maximilian Dörrbecker, Wikimedia Commons, CC BY-SA 2.5)

However, the term "the" Great Wall of China is a misnomer. In fact, there is a series of overlapping walled fortifications that began being constructed by early dynasties in the 5th century BCE and continued through to the 17th century CE. Walls are a defensive military structure and are thus an expression of a civilization that wished to be left alone. Emperors generally disregarded China's extensive coastline, and where port cities did emerge, they were primarily used for local trade.

In Southeast Asia, however, trade links with South Asia brought Hinduism and later Buddhism to the region. Port cities emerged, as well as cities that were religious or ceremonial centers. The

Hindu rulers of the region were often viewed as divine, but in order to secure the favor of the gods, and the blessings of the Hindu priests, they agreed to build temples. Angkor Wat, in Cambodia, for instance, was built in the 12th century as the king's state temple and capital city (see **Figure 9.7**). It was later transformed into a Buddhist temple, which it remains today. The temple complex is the largest religious structure in the world.

Figure 9.7: Angkor Wat Temple Complex, Cambodia (© Bjøorn Christian Tøorrissen, Wikimedia Commons, CC BY-SA 4.0)

Eventually, Islam spread to Southeast Asia, particularly as a result of Sufi missionaries, part of a mystical branch of Islam. In the present-day islands of Malaysia and Indonesia, in particular, local rulers and communities embraced Islamic theology. Today, more Muslims live in Indonesia than in any other country on Earth.

Buddhism continued to dominate the religious landscape of much of Southeast Asia as well as in Japan. During the Heian period, lasting from the late 8th centuries to the 12th century CE, many of the features of modern Japanese culture emerged, such as its distinctive art and poetry, as well as Buddhist-inspired architecture. A ruling class of warriors, known as a shogunate, would later take control of Japan beginning a feudal period in the country's history.

The evolving landscape of this region would be completely transformed by colonization, with sweeping political and economic changes that continues to shape the geography of the region today. Beginning in the 16th century, European colonial empires became interested in Southeast Asia. Before long, Europeans established permanent colonies. The Spanish would settle the

Philippines, the Netherlands established the Dutch East Indies in present-day Indonesia, the French created Indochina in mainland Southeast Asia, and the British would take over Burma, now known as Myanmar, and Malaysia. By the 1800s, only Thailand would remain independent and functioned largely as a buffer state separating the British and French colonial spheres.

Japan took note of these imperial pursuits. In 1868 CE, the Japanese Emperor Meiji ended the shogunate and began a series of reforms known as the Meiji Restoration. As part of the reform, the government sought to increase Japan's modernization and industrialization and began a systematic study of the developed world. Why were some countries more powerful and more industrialized than others? Britain, for example, was an island nation like Japan and yet was considered to be the most powerful country in the world. Education was critical, as was industrial technology, but Japanese leaders believed that Britain's colonial ambitions, its direct control over the resources of other areas, was key to its success.

By the beginning of World War II, Japan had built up an impressive military and had colonized much of East and Southeast Asia including northeastern China, the Korean peninsula, Taiwan, French Indochina, the Philippines, Indonesia, and Malaysia (see **Figure 9.8**). In 1941, Japanese military forces attacked the US base Pearl Harbor in Hawaii. Following the attack, the US declared war on Japan and entered World War II.

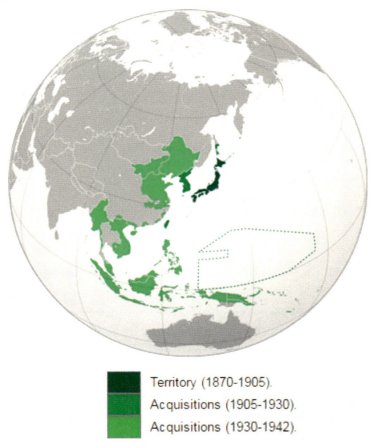

Figure 9.8: Map of the Japanese Empire, 1942 (Derivative work from original by Shadowxfox, Wikimedia Commons, CC BY-SA 3.0)

Following Japan's loss in World War II, the countries of East and Southeast Asia were able to acquire independence. Some countries, like the Philippines and Burma, achieved independence through a peaceful turnover of control, while others such as Indonesia won independence only after a violent period of opposition. The end of World War II reshaped not only the political map of East and Southeast Asia but development in the region as well.

9.4 POLITICAL CONFLICTS AND CHANGES EAST AND SOUTHEAST ASIA

The second half of the 20th century was a time of significant political change for East and Southeast Asia. The former colonies of Japan were able to break away from their colonial past and become independent, but as in many other parts of the world, that independence often coincided with political conflict.

For Japan, the end of World War II brought a period of **Westernization** and rapid economic growth. Westernization refers to the process of adopting Western, particularly European and American, culture and values. Japan adopted a new constitution and embraced democratic principles. It continued to industrialize and would become a global leader in electronics and automotive production. Today, Japan has the fourth largest GDP behind only the United States, the European Union, and China.

In other parts of East and Southeast Asia, the political changes to the region following WorldWar II tended toward **communism**, a social, political, and economic system that seeks communal ownership of the means of production. Communism is associated with **Marxism**, an analysis of social class and conflict based on the work of Karl Marx (1818-1883 CE). In a typical society, factories are owned by a wealthy few who then pay workers a lower wage to ensure that they make a profit. In a communist society, however, the goal of Marxism would be a classless society where everyone shares the ownership and thus receives equal profits.

Marxist ideas spread to China by the early 20th century and found particular support among Chinese intellectuals. The Communist Revolution in Russia inspired Marxists in China who founded a communist political party that would eventually be led by Mao Zedong. The communist party continued to gain traction in China and following a civil war, Mao Zedong established the communist People's Republic of China in 1949. The previous Chinese government fled to the island of Taiwan, which is officially known as the Republic of China and claims control of the entire mainland. China, however, maintains that Taiwan is part of China.

After securing political control of China, Mao Zedong sought to transform China's culture by reorienting it around the ideology of communism. One of the first steps in this transformation was the **Great Leap Forward** from 1958 to 1961 which sought to reshape China's agrarian society into an industrial power. Unfortunately, the changes led to widespread famine and the deaths of tens of millions of Chinese as a direct result.

Following the failure of the Great Leap Forward, Mao aimed to eliminate any remaining traditional elements of Chinese culture or capitalist thinking through the Cultural Revolution. Millions were imprisoned, forcibly relocated, or tortured, and historical relics and cultural sites were destroyed. After Mao's death, several leaders responsible for the abuses committed during the Cultural Revolution were arrested and China began a period of modernization and economic reform.

In the Korean peninsula, allied forces divided the former Japanese colony along the 38th parallel. Russia would control the northern portion, where it helped install a communist government and economic system. The United States occupied the southern portion, where it assisted a pro-Western government in its political and economic development. Tensions between the two territories led to the Korean War in the early 1950s. Technically, the two sides are still at war having never signed a peace agreement and simply agreeing to a cease-fire. Today, North Korea, officially the Democratic People's Republic of Korea (DPRK), follows a Marxist model of development with state owned enterprises and agriculture. The government has been accused of numerous human rights violations and the people of North Korea are severely restricted in terms of their economic, political, and personal freedom. In South Korea, on the other hand, officially known as the Republic of Korea, a democratic government replaced a series of military dictatorships and the country is considered one of the most developed in the region according to the Human Development Index.

Communist ideals spread to Southeast Asia, as well, where Marxism influenced the governments of newly independent countries. In Vietnam, for example, a communist movement was begun by Ho Chi Minh to try to gain independence from France following the end of Japanese occupation in WorldWar II. The communist forces were able to defeat the French in a key battle in 1954 and established a government in the northern territory. The country was then divided into a communist north and anti-communist and majority Catholic south. This was a time of high tension between the United States and the Soviet Union, and the US feared that the entire region would eventually come under communist control, essentially creating a Western capitalist hemisphere and an Eastern communist hemisphere. The fear that the fall of one country to communism would lead to the fall of other surrounding countries to communism was known as **domino theory**, and was originally meant as an anecdote but became the basis for US foreign policy in the region (see **Figure 9.9**).

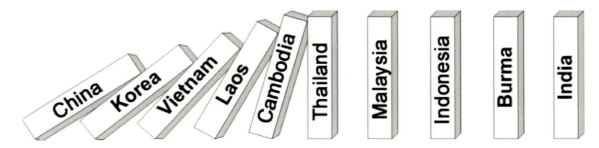

Figure 9.9: Illustration of Domino Theory (© User:Nyenyec, Wikimedia Commons, CC BY-SA 3.0)

The United States aimed to support South Vietnam's resistance to the communist north's goal of unification and began sending military advisors to the region. Military combat units followed and bombing campaigns began in 1965. The terrain of Vietnam was quite different than the geography of other areas where the US had previously fought. Much of Southeast Asia was tropical rainforest, and was ill-suited for the types of tanks and heavy artillery that had been so successful in World War II. The Viet Cong, referring to the Vietnamese communists, engaged in guerrilla warfare, using the terrain to support small, mobile military units. To try to combat these tactics, the US military sprayed chemical defoliants and herbicides, like Agent Orange, over Vietnam's forests. In the end, waning support for the Vietnam War led the US to withdraw and in 1975, Vietnam was unified under communist rule. Over 1 million Vietnamese and 58,000 Americans died in the fighting. Millions others were exposed to Agent Orange causing health problems and disabilities, and the chemical had devastating effects on Vietnam's ecosystem where it has lingered in the soil.

During the same time period, a communist organization known as the **Khmer Rouge**, which is French for "Red Khmers," came to power in Cambodia. Khmer refers to the dominant ethnic group in Cambodia. The Khmer Rouge opposed Westernization and US involvement in the newly independent country and believed in a return to an agrarian society. Pol Pot (1925-1998 CE), the leader of the Khmer Rouge, led a campaign to eliminate the country's schools, hospitals, and other institutions and make the entire society work on collective farms. Urban cities would no longer be the economic and political focus, but rather wealth would be spread out around the countryside. Most of the country's intellectuals, including teachers and even people with glasses who were simply perceived as academic, were killed. Large prison camps were set up to house those who were believed to be a threat to communism. Cambodians of other ethnicities or who practiced religion were also executed. In total, more than one million people were killed, often buried in mass graves known as the Killing Fields. Cambodia's attempt to transform into an agrarian society ultimately led to widespread famine and starvation. In 1978, Vietnamese forces invaded Cambodia and defeated the Khmer Rouge, but human rights continue to be severely restricted in the country.

Much of East and Southeast Asia exhibits characteristics of a **shatter belt**, an area of political instability that is caught between the interests of competing states. Beginning in the colonial era and continuing today, Western involvement in this region has at times led to industrialization and economic growth and at other times economic depression and a drive to return to traditional values. Today, political instability continues to plague several countries in the region.

9.5 PATTERNS OF ECONOMIC DEVELOPMENT IN EAST AND SOUTHEAST ASIA

Despite the political changes and conflicts that marked the 20th century, the 21st century has largely been marked by economic development across East and Southeast Asia. Economic geography is a branch of geography that explores the spatial aspect of economic development. Economic geographers don't just ask "Where is economic development occurring?" but also *"Why is economic development occurring in some areas and not others?"* In East and Southeast Asia, economic growth has largely resulted from regional and global trade. However, development is not spread evenly across the region and economic inequalities still persist.

Global connections have been the principle driver of economic success in East and Southeast Asia. Much of the trade connections have emerged between the countries in this region and the larger **Pacific Rim**, referring to the countries that border the Pacific Ocean. Many of these countries are members of the Asia-Pacific Economic Cooperation (APEC) which promotes free trade across Asia and the Pacific. In Southeast Asia, the countries of the region formed the

Association of Southeast Asian Nations, or ASEAN (see **Figure 9.10**). The organization aimed to promote political security, economic growth, and social development among member countries.

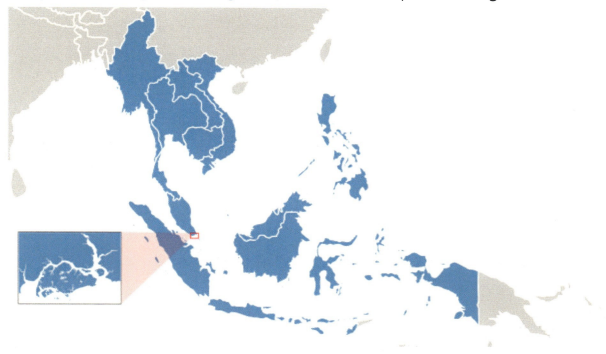

Figure 9.10: Map of ASEAN Member States (Derivative work from original by ASDFGH, Wikimedia Commons, Public Domain)

In China and Japan, histories of relative isolation gave way to an embrace of globalization and global trade. Although China's government is communist, it has allowed more free-market oriented economics in areas known as **Special Economic Zones**, or SEZs. These SEZs are located in coastal China and have special incentives to attract foreign investment (see **Figure 9.11**). Other capitalist shifts have occurred in China even allowing for US supermarkets and restaurants to open locations in the country. These SEZs, as well as other special development areas in China, have functioned as **growth poles**, which are areas that have attracted economic development in the region. In 2010, China displaced the United States as the global leader in manufacturing.

Figure 9.11: Map of China's Special Economic Zones (© Alan Mak, Wikimedia Commons, CC BY 2.5)

Broadly, **foreign direct investment** (also called FDI), the control of a business in one country by a company based in another country, has been a key driver of China's economic success. In 2017, China was the second most attractive company for foreign investors, behind the United States, with over $136 billion in foreign direct investment flowing into the country according to the 2018 World Investment Report. China has done its own foreign direct investing as well, increasing outward flows of FDI from $5.5 billion in 2004 to over $125 billion in 2017, according to the same report. Hong Kong is the primary destination for Chinese foreign direct investment, but substantial sums also flow to countries in Africa as well as Australia, where China is the largest trading partner. Within Southeast Asia, China is now the top investor in Myanmar and has increased foreign direct investment in Singapore.

While China and Japan remain the largest economies in the region, other areas have experienced high rates of economic growth. The Four **Asian Tigers**, in particular, referring to Hong Kong, Singapore, South Korea, and Taiwan, experienced rapid industrialization and economic development led by export-driven economies, low taxes, and free trade. Some have also theorized that the Confucian work ethic present in these countries complemented the process of industrialization. Each country also developed a distinct specialty and maintains a competitive advantage in that area. South Korea, for example, is known for its manufacturing of information technology while Hong Kong is a leading financial center.

Much of the economic growth of the Asian Tigers as well as Japan has come from the export of **value added goods**. When countries export raw materials, their economic benefit is limited since those raw materials are often not inherently valuable. When these raw materials are made into something useful, however, value is added and the product can be sold for a higher profit. Lumber is quite cheap, for example. When the lumber is made into a dining table, it has a much higher value. Many companies located in these countries have become household names in electronics, computing, and the auto industry. China has traditionally exported relatively low value added goods, but in 2016, the Chinese government announced a shift to higher value added products like transportation technology and telecommunications.

Southeast Asia has benefitted from its key geographical position as the *entrepôt*, a French term meaning a commercial center of trade, for the rest of Asia. The Strait of Malacca in particular is the main shipping channel between the Pacific Ocean and the Indian Ocean and a key transportation gateway (see **Figure 9.12**). Around one quarter of all the world's exported goods travels through the strait each year. Malaysia's economic success as an entrepôt is exemplified by its Petronas Towers in Kuala Lumpur, which were the tallest buildings in the world from 1998 to 2004. Indonesia has the largest economy of Southeast Asia, exporting primarily to Japan, Singapore, the United States, and China. Singapore has the highest GDP per capita in the region, again owing to its key geographical position.

Figure 9.12: Map of the Strait of Malacca (U.S. Department of Defense, Public Domain)

One lingering issue in many of these countries has been crony capitalism, the notion that the success of a business depends on its relationship to other businesses and the state. A politician might have an old friend in the manufacturing industry, for example, and give the friend a government contract with beneficial terms. In 1997, a financial crisis that started in Thailand spread throughout the Southeast Asia region and to South Korea was, in part, blamed on the business dealings of corrupt politicians. Several countries in this region rank high on an index of corruption perception, as shown in **Figure 9.13**, and some have expressed concern that the communist countries of this region will continue to embrace capitalism when it is politically beneficial rather than as part of a broader and more transparent economic system.

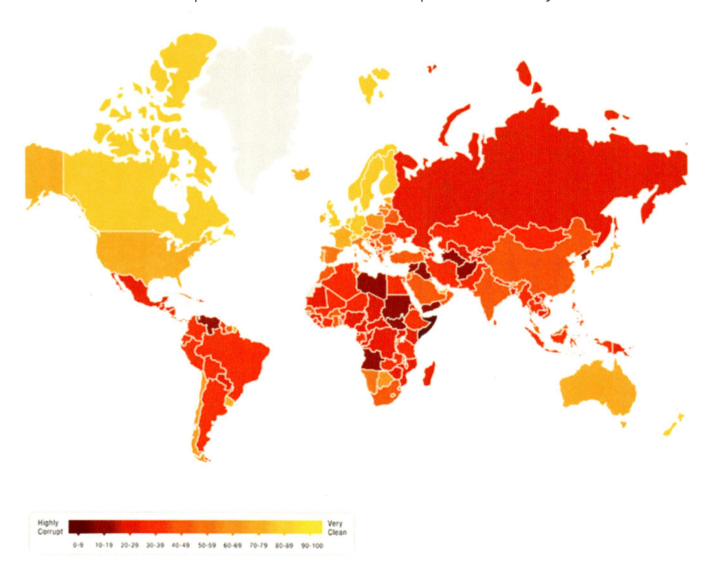

Figure 9.13: Map of Index of Perception of Corruption by Transparency International, 2017 (© Transparency International, CC-BY-4.0-DE)

Economic development is not spread evenly throughout the region. By most estimates, North Korea remains the poorest country in the region with an estimated GDP per capita of just $1,300 in 2016. Mainland Southeast Asia, with the exception of Thailand, remains relatively poor. Cambodia, Laos, Myanmar, and Vietnam all have GDP per capitas of less than $8,000 as of 2018 and remain far less well off than their more developed neighbors in the region.

Where economic growth has occurred, it is often confined to an urban area which can drive up population densities as people move to the city in search of work. In Indonesia, the island of Java is very densely populated while other surrounding islands are relatively sparse. The Dutch colonizers and later the Indonesian government, in a stated effort to reduce poverty and overcrowding, created a policy of transmigration seeking to relocate people to the less densely populated islands. This program has been controversial, however, and the indigenous people who inhabit the surrounding islands see the program as a threat to their way of life.

In some cases, uneven distribution of economic development has led to both interregional and intraregional migration as people move in search of economic advancement. China in particular has seen a significant amount of rural to urban migration. Around 11 percent of the entire country's population migrated from rural to urban areas in 2009 and most of them are young adults. Many of China's migrants are part of a **floating population**, which refers to members of a population who reside in an area for a period of time but do not live there permanently. Around 50 million Chinese reside overseas, mostly in Southeast Asia. Thailand has the largest population of overseas Chinese, and Chinese also represent the majority ethnic group in Singapore. Long before colonization, East and Southeast Asia was a realm of global economic influence, from the Chinese empire to the trade routes of Southeast Asia. As some countries of the realm have moved toward political stability and economic growth, others have remained in a state of political and economic turmoil. Tourism could bring some of these countries an economic boost, but the prospect of tourism in countries with government instability is limited. Still, although countries like Laos, Myanmar, and Cambodia currently have some of the world's smallest economies, they are some of the fastest growing in the world and improvements in agricultural and natural resource development combined with a stable political infrastructure could expand the economic strength of the region.

CHAPTER 10

Oceania

> **Learning Objectives**
>
> - Identify the key geographic features of Australia, the Pacific, and the polar regions
> - Describe the biodiversity found in Australia and the Pacific
> - Explain the patterns of human settlement in Oceania
> - Analyze how climate change is impacting the geography of Oceania

10.1 THE PHYSICAL LANDSCAPE OF OCEANIA

Oceania is a realm like no other. Nowhere else in the world can one find some of the unique wildlife that is found in this realm, and no other region is as isolated. Oceania is the only world region not connected by land to another region. This is a region of the world at a crossroads where the effects of global changes in climate and pollution could have profound effects. The region of Oceania includes Australia, the realms of the Pacific Islands, and the polar regions of the Arctic and the Antarctic. While some regions share a distinct cultural or colonial history and others share a common physical landscape, the region of Oceania is connected more by its isolation than by a shared physiography or human experience.

Australia dominates the region in terms of size, economics, and population. The country has the unique designation of being both a sovereign state and a continent. Often, Australia and New Zealand are considered a single region (see **Figure 10.1**), but while the two countries share cultural and historical similarities, their physical landscapes are quite different. Australia lies in the middle of its own tectonic plate making it relatively geologically stable. Australia has no active volcanoes and has had only a small number of large earthquakes. Its tectonic position also limits its relief and much of the continent is relatively flat. An exception to this is the Great Dividing Range which runs along the coast of Eastern Australia. This series of mountain ranges affects Australia's climate by providing orographic rainfall along the coast and divides the core population center of Australia from the rest of the continent.

Figure 10.1: Physical Map of Australia and New Zealand (CIA World Factbook, Public Domain)

The other key geographic feature of Australia is its vast interior known as the **Outback** (see **Figure 10.2**). This remote area of extensive grassland pastures supports one of the world's largest sheep and cattle industries. However, the ecosystem of the Outback is quite fragile. With limited precipitation and vegetation, overgrazing puts the region at risk for desertification. In addition, although this region was the center of population for Australia's indigenous groups, ranching in the Outback has created issues of land ownership.

Figure 10.2: View of Australian Outback and Mount Conner (© Gabriele Delhey, Wikimedia Commons, CC BY-SA 3.0)

One of the most well-known features of Australia's geography lies just off coast: the **Great Barrier Reef**. This massive underwater reef is the world's largest coral structure and stretches over 2,300 kilometers (1,400 miles). However, warming ocean temperatures and pollution have been a significant environmental threat to the Great Barrier Reef in recent years.

Unlike its geologically stable neighbor, New Zealand is situated at the intersection of the Pacific Plate and the Australian Plate (see **Figure 10.3**). Its two large, mountainous islands and numerous small islands are prone to both earthquakes and volcanoes. New Zealand is younger than Australia geologically and has a far more varied topography. On New Zealand's North Island alone, you could spend the morning surfing on a sandy beach, the afternoon picnicking in the rolling green hills where the fictional city of Hobbiton was filmed, and the evening skiing on an active volcano, Mount Ruapehu. New Zealand's South Island is home to a number of stunning fiords, more commonly found in Scandinavia where they are spelled fjord.

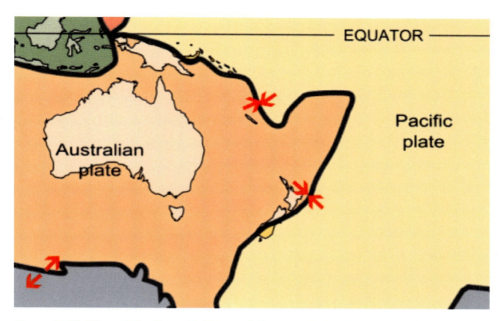

Figure 10.3: Map of the Tectonic Plates of Oceania (United States Geological Survey, Public Domain)

The islands of the Pacific to the north and east of Australia and New Zealand are divided into three regions (see **Figure 10.4**). New Zealand is part of the islands of **Polynesia**, from the prefix "poly" meaning "many." Polynesia is a large, triangular region stretching from New Zealand to Easter Island to the Hawaiian and Midway Islands. West of Polynesia and to the northeast of Australia are the islands of **Melanesia**, including New Guinea, the Solomon Islands, and Fiji. Europeans called the region "Melanesia" from the Greek prefix melan- meaning "black," referring to the darker skin they believed characterized the people of this realm. North of Melanesia are the tiny islands of **Micronesia**, from the prefix "micro" meaning "small." There are over 2,000 islands in Micronesia.

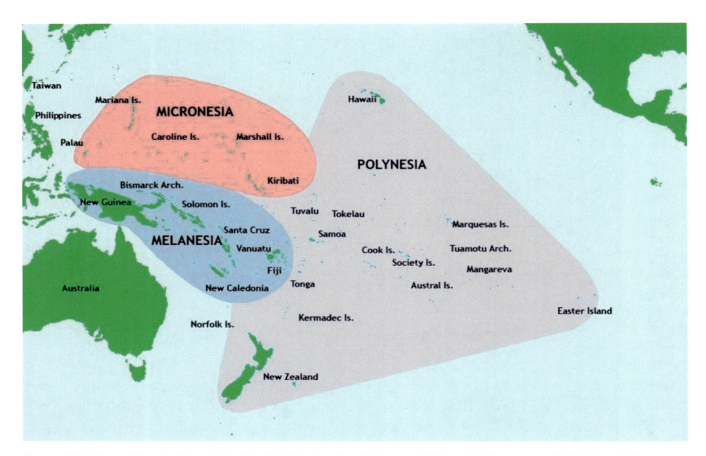

Figure 10.4: Map of Australia and the Pacific (Map by User:Kahuroa, Wikimedia Commons, Public Domain)

The islands of the Pacific can be divided into two groups based on their physical characteristics. The **high islands** like Hawaii are volcanic, meaning they were formed from volcanoes, and thus have a relatively high relief. This high relief and volcanic soils enables the high islands to have fertile soil and ample rainfall, which in turn supports a diverse agricultural system and relatively large populations.

In contrast, the **low islands** of the Pacific are comprised mostly of coral and, as their name implies, are generally low in elevation. Most of the islands in the Pacific, particularly in Micronesia, are low islands. These islands may only rise a few feet above the water and their dry, sandy soil makes farming difficult. Fresh water is often in short supply in the low islands. As a result, these islands typically have much smaller populations. The relatively large coral island that comprises the country of Niue, for example, rises to a maximum 60 meters (less than 200 feet). The low elevation of these islands make them vulnerable to natural disasters, such as tropical cyclones, and to changes in sea elevation due to rising global temperatures. In the country of Tuvalu, an island chain located between Hawaii and Australia, the highest point is a mere 4.6 meters (15 feet) above sea level and the island has sustained severe damage from cyclones during its history.

A number of low islands of in the Pacific form **atolls**, ring-shaped chains of coral islands surrounding a central lagoon (see **Figure 10.5**). Typically, the lagoon is actually a volcanic crater which has eroded beneath the water. Most of the world's atolls are found in the Pacific Ocean and their land areas are generally quite small.

Figure 10.5: Satellite Photo of the Atafu Atoll in Tokelau (NASA Johnson Space Center, Public Domain)

In general, the islands of the Pacific have warm, tropical climates with little seasonal extremes in temperature. Some islands experience seasonal, primarily orographic rainfall. These relatively warm temperatures help support tourism throughout the region. Throughout New Zealand and the core area of Australia, east of the Great Dividing Range, is primarily a **maritime climate**. This climate zone features cool summers and winters with few extremes in temperature or in rainfall.

Also included in Oceania are the earth's polar regions. In the North Pole is the Arctic Ocean, the world's smallest and shallowest ocean. Although it may appear to look like a landmass covered in snow on many globes, there is no landmass below the North Pole. The ocean is covered by a sheet of sea ice throughout the year and the entire body of water is almost completely ice-covered in winter. In the South Pole is Earth's southernmost continent, Antarctica. This continent is around twice the size of Australia and is almost entirely covered with ice. It is not home to a permanent human settlement.

10.2 THE WORLD'S OCEANS AND POLAR FRONTIERS

Over 70 percent of the entire surface of the world is covered with water, but who controls it? If the body of water is inland, ownership is quite clear. A lake in the interior of a state belongs to that state. For the 96.5 percent of the world's water that's held in oceans, however, ownership is much less clear. Historically, the world's oceans were considered the "high seas" and while states had control over their immediate coastline extending out three miles, the vast stretches high seas were free from ownership. As ocean resources became more important, however, countries became interested in establishing clear rights to minerals, oil, and fishing stocks offshore.

In 1945, President Harry S. Truman announced that the sovereign territory of the United States extended to the boundary of its continental shelf, which was in some places hundreds of miles offshore. Other countries, including Chile, Peru, and Ecuador, followed suit, beginning an international dash to claim offshore waters. Within two decades, countries were using a variety of systems of ownership; some claimed waters three miles offshore, others 12 miles, and still others maintained ownership over all of the waters to the continental shelf.

Eventually, the United Nations intervened, seeking a universal system of ocean ownership. The United Nations Convention on the Law of the Sea (UNCLOS) resulted from series of international conferences and established guidelines for maritime travel and control of natural resources found in the world's seas. As a result of the UNCLOS, there are now several categories of ownership over the world's water depending on its distance from shore (see **Figure 10.6**). A state's internal waters are considered the sovereign territory of a state. **Territorial waters** extended 12 miles offshore and are also considered sovereign territory of a state. However, in territorial waters, a state must grant "innocent passage" to oceangoing vessels, meaning it must allow the vessel to pass through as long as it is doing so in a speedy manner that is not threatening the security of a state. Beyond the territorial waters, a state can control certain aspects of a 12 mile contiguous zone, including taxation and pollution. Following the US claim of control over the continental shelf, the UNCLOS established that a 200 mile zone extending out from a country's coastline was its **exclusive economic zone**, or EEZ, where it has exclusive control over any natural resources. Other countries can fly over or pass through the waters of the EEZ, but cannot use the resources within. However, countries are free to sell, lease, or share the rights to their EEZ. Beyond the EEZ are international waters where no state has direct control.

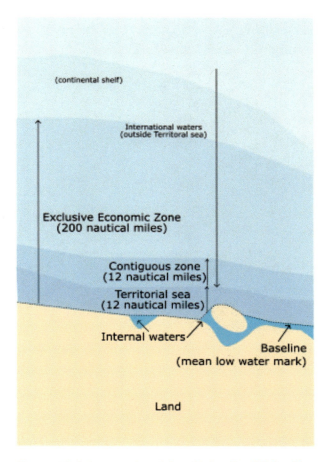

Figure 10.6: International Sea Rights Established by the UNCLOS (© historicair, Wikimedia Commons, CC BY-SA 3.0)

As a result of the UNCLOS, some tiny islands gained immense stretches of ocean territory – and the rights to the resources in and underneath those waters (see **Figure 10.7**). Some countries found this as an opportunity to expand their resource area. Conflicts developed over what would otherwise be tiny specks of island territory but what had become over 100,000 nautical miles of ocean resources. Particularly as the technology for offshore drilling improved, states sought to secure control of what could be huge caches of oil and minerals.

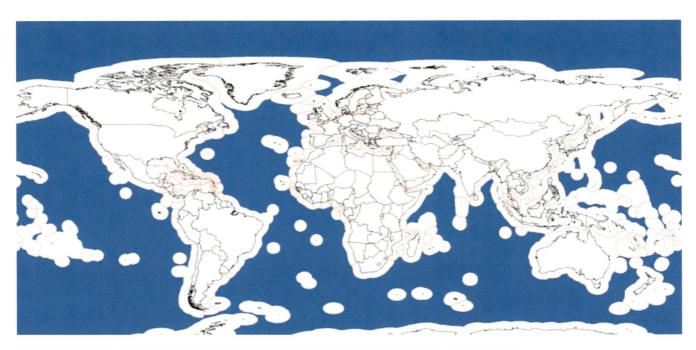
Figure 10.7: Map of International Waters (© Kvasir, Wikimedia Commons, CC BY-SA 3.0)

The UNCLOS also established some ownership over the Arctic Ocean. Russia, Norway, Canada, the United States, and Denmark, which controls Greenland, all have overlapping EEZs in the Arctic. Historically, this frigid, isolated region was of little interest to countries. Early attempts at exploration were largely unsuccessful and a person wouldn't reach the North Pole until the early 20th century. However, the drive to secure fossil fuels has led to more intensive research and exploration in the region and as much as one-quarter of the entire world's oil and natural gas reserves are believed to lie below Arctic waters. Global increases in temperature could further open up previously inaccessible areas of the Arctic to drilling operations. A 2015 declaration signed by all five states surrounding the Arctic prohibited fishing in the central Arctic Ocean in an effort to protect ocean life and resources.

In the South Pole, Antarctica remains a frontier region with no permanent human inhabitants, though the continent is home to penguins, fur seals, and other marine creatures. Antarctica does have a number of research stations as well as an Orthodox Church and a few thousand people work in and around Antarctica in various times of the year conducting scientific research. Antarctica is the coldest place in the world, once dropping down to -89.2 °C (-128.6 °F) at a Russian research station. Although Antarctica might look relatively moist and snow covered, it is actually a desert with very little precipitation.

So who controls this vast expanse of frozen desert? The answer depends on who you ask. Several different countries claim control of Antarctic territories (see **Figure 10.8**) but, in general, these states do not recognize each other's claims.

OCEANIA

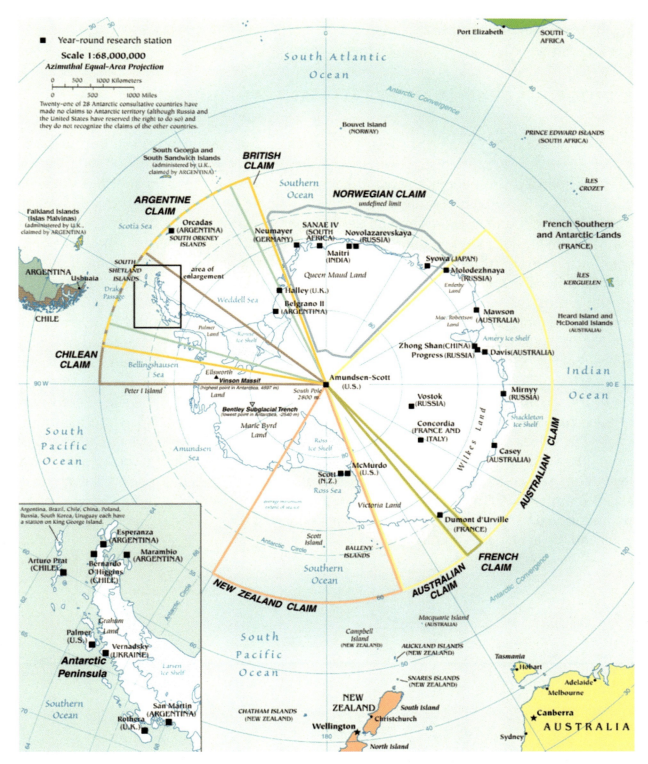

Figure 10.8: Map of Research Stations and Antarctic Territorial Claims (CIA World Factbook, Public Domain)

In 1959, the Antarctic Treaty was signed that put a hold on new territorial claims, established Antarctica as a zone for scientific research and environmental protection, and prohibited military activity in the region. A later treaty signed in 1998 reaffirmed Antarctica as a peaceful, scientific frontier and prohibited mining on the continent.

10.3 BIOGEOGRAPHY IN AUSTRALIA AND THE PACIFIC

The relative isolation of Oceania defines it as a region but has also contributed to perhaps its quirkiest characteristic: its distinctive wildlife. **Biogeography** is a branch of geography that explores the spatial distribution of the world's flora (plant life) and fauna (animal life). While every world region has its own, unique plants and animals, some of the creatures found in Australia and the Pacific are found nowhere else on Earth. A number of world regions have an impressive degree of **biodiversity**, meaning there are a wide variety of species present. These regions are generally located in the tropics. British biogeographer Alfred Russel Wallace was one of the first to try and determine the boundary of Australia and Southeast Asia's unique plants and animals in the 19th century.

Geography is more than the *where*, however, but is also a discipline that asks "Why?" *Why* does Australia have such unique, and frankly a bit frightening at times, flora and fauna? *Why* are **monotremes**, mammals that lay eggs rather than give birth to live young, only found in the isolated region of Australia and New Guinea? It is the isolation of this region that's key. 200 million years ago, Australia was situated on the far-reaches of Pangaea, the last supercontinent (see **Figure 10.9**). Around 175 million years ago, Pangaea began to break apart. During the same time period, the earliest mammals were evolving, diverging first from egg-laying reptiles and then continuing to adapt and change. Early egg-laying monotremes were found throughout Pangaea but eventually went extinct in the other world regions, out-competed by more evolutionarily advanced mammals. Australia and New Guinea, however, broke away before more advanced mammals arrived, and thus monotremes remained. The only modern monotremes are the platypus and the echidna.

Figure 10.9: Map of Pangaea with Modern Continental Outlines (© User:Kieff, Wikimedia Commons, CC BY-SA 3.0)

In addition to monotremes, Australia is home to the world's largest and most diverse array of marsupials, mammals who carry their young in a pouch. A number of marsupials are also found in Central and South America and just one species lives in North America: the Virginia opossum, more typically referred to as a "possum." In Oceania, well-known marsupials include the kangaroo, koala, wombat, and the Tasmanian devil. The Kangaroo in particular is a widely used symbol of Australia and kangaroo meat, though controversial, can be found throughout Australia.

Australia is not just home to cuddly marsupials like koalas, wallabies, and quokkas, though. It is also home to some of the world's deadliest creatures. There are more deadly snakes in Australia than in any other country in the world including the taipan, considered by some to be the world's most venomous. An episode of the children's television show "Peppa Pig" was actually banned in Australia because it featured a "friendly spider" and local officials believed it would send the wrong message to Australian children in a country where spiders can be deadly. Offshore, Australia's box jellyfish can kill simply by the pain inflicted by its sting which can send the body into shock.

Another key area of biodiversity is New Zealand, particularly in terms of its flora. Its isolation allowed for species of trees that have remained relatively unchanged for the past 190 million years. Several species of birds in New Zealand, such as the moas, went extinct due to hunting shortly after humans first arrived in the region. Few mammals existed in New Zealand before human settlement and the arrival of the first mammals here, such as rats and weasels, led to widespread extinctions of native species that had never had to evolve to compete with these predators. Rabbits in particular proved to be a particularly troublesome **invasive species**, which refers to a species of plant, animal, or fungus that is not native to an area but spreads rapidly. Early settlers to New Zealand brought rabbits for fur and meat, but the high reproductive capacity of rabbits quickly proved troublesome and by the 1880s, rabbits were having a considerable negative effect on agriculture. The solution in the late 19th century was to introduce stoats, ferrets, and weasels, natural predators of rabbits. Unfortunately, these species proved devastating to local bird species and had only minimal impact on the increasing rabbit population. Cats were similarly introduced, but they too caused the extinction of several bird species and a native bat. (The children's story about the old lady who swallowed the fly comes to mind.) These species are still seen as some of the biggest threats to New Zealand's wildlife.

10.4 THE PATTERNS OF HUMAN SETTLEMENT IN AUSTRALIA AND THE PACIFIC

Much of the physical landscape of Oceania has been directly shaped by human activity and settlement. Although Australia today is known for its origin as a British prison colony, the continent was inhabited long before Europeans arrived. The indigenous people of Australia are known as **Aborigines** and comprise a number of different ethnolinguistic and cultural groups. Most researchers believe the first aboriginal groups arrived in Australia between 40,000 and 50,000 years ago when sea levels were lower and land bridges and relatively short sea crossings separated Australia, Tasmania, and Papua New Guinea from mainland Southeast Asia.

It took thousands more years and advances in ocean transportation and navigation for the rest of the Pacific islands to be settled (see **Figure 10.10**). Humans gradually made their way to the islands of Melanesia, to Fiji by 900 BCE then east and north. The far-reaches of Polynesia, including Hawaii and Easter Island, were not populated until much later due to the long distances separating them from other landmasses. New Zealand, though, was one of the last to be settled, with Eastern Polynesians not arriving on the islands until around 1250 CE. These groups developed their own ethnic and cultural identity known as the **Maori**.

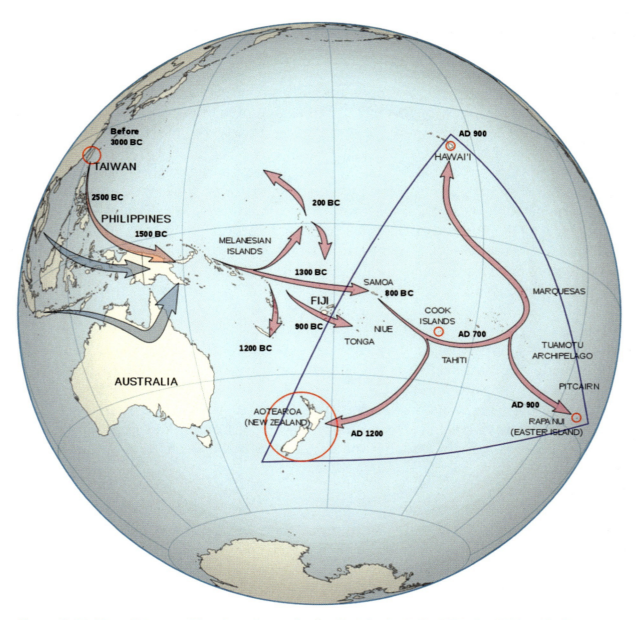

Figure 10.10: Map of Human Migrations Across the Pacific Islands (© David Eccles, Wikimedia Commons, CC BY 4.0)

Life would change dramatically for the people of Oceania with the arrival of Europeans. The Dutch first made landfall in Australia in 1606 CE but simply explored and mapped the area and did not establish a settlement. In the late 18th century, the British established their first Australian settlement in what would later become the city of Sydney with the intention of creating an overseas penal colony. However, many of the prisoners sent to Australia were not hardened criminals who needed to be separated from the British Isles by an expansive ocean. Many were accused of petty crimes like theft and even children who had committed crimes were shipped to Australia. Today, around 20 percent of Australians are the descendants of these imprisoned settlers.

European settlement of Australia challenged aboriginal land and water resources, but it was disease that had the most devastating effect on the indigenous population. At the time of British colonization, there were likely between 315,000 and 750,000 Aborigines in Australia. By the start of World War II, diseases like smallpox and measles reduced their numbers to just 74,000.

New Zealand was originally claimed by the British as a colony of Australia, but then became its own colony in the mid-19th century. Around the same time, representatives of Britain as well as Maori leaders signed the Treaty of Waitangi. This treaty granted British colonists sovereignty over the governing of New Zealand but gave the Maori the rights to their tribal lands and resources and made them British subjects.

Throughout the 19th and 20th centuries, European and Japanese colonial expeditions claimed most of the Pacific islands. Some islands were seen as strategic military bases. Others, such as France's colony of New Caledonia, were transformed into overseas prison colonies following the British model. Still others were occupied for their resources and trade opportunities. In the decades following World War II, a number of islands achieved independence. Australia slowly increased its autonomy throughout the early 20th century, officially dissolving from British control in 1942. New Zealand gained independence from Britain in 1947. In the 1970s and 1980s, another wave of independence occurred, with Fiji, Tonga, and a number of other states gaining independence.

Others were not granted independence and a number of Pacific islands remain colonies today. Hawaii was made a US state in 1959 largely against the wishes of its indigenous population. Guam became a strategic US Naval base in the Spanish-American War and later in World War II and remains a US territory today. Its residents are US citizens but cannot vote in elections. The vast majority of the world's remaining colonies today are islands because of their strategic locations and resource potential, particularly after the United Nations Convention on the Law of the Sea (see **Figure 10.11**).

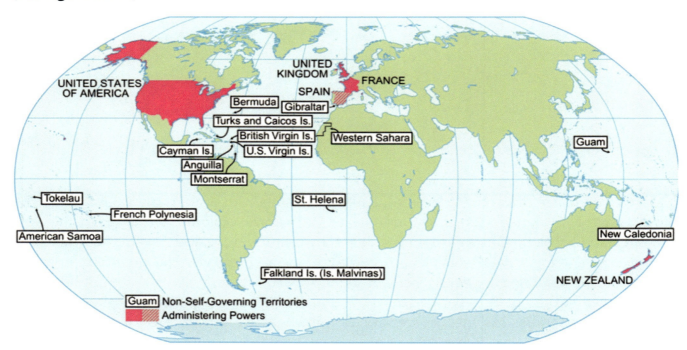

Figure 10.11: Map of Non-Self-Governing Territories, 2012 (Derived from UN Map of Non-Self Governing Territories, Wikimedia Commons, Public Domain)

Although Australia and New Zealand gained independence relatively early compared to many of the other areas of Oceania, these countries have experienced lingering issues related to their indigenous populations. For Australian Aboriginals, there were two historical issues they sought to address: the government admission of mistreatment and ownership over tribal lands. Previously, the Australian government ruled that Australia prior to British arrival was considered *terra nullius*, or nobody's land, and thus there was no regard for indigenous rights to land or resources. In 1992, however, Australia's high court ruled that land policy was invalid. By this time, Australia's Northern Territory was already considered aboriginal land, but later courts found that over three-quarters of the land in Australia could be subject to aboriginal claims, even though Aborigines comprised only 3 percent of the population of Australia. In 2008, the prime minister of Australia issued a formal apology to the Aboriginal people of Australia on behalf of the government.

One challenge to these perhaps promising legal developments, however, is that private enterprise is prohibited on aboriginal land. Even building a home is seen as a violation of the current guidelines. Thus, many Aborigines became land-rich as a result of the government's decision, but still remained poor. Unemployment, alcohol abuse, food security, and land reform continue to be significant issues for Australian Aborigines. Today, there are over 500,000 Aboriginal Australians and almost one-third now live in Australia's major cities.

The Maori of New Zealand make up a much larger portion of the country's population at around 15 percent. The Maori have generally kept their traditional cultural and linguistic traditions while partially integrating into more western New Zealand society. Compared to other groups in New Zealand, the Maori have lower life expectancies and average incomes, and make up around 50 percent of New Zealand's prison population.

Economically, Australia and the Pacific islands have struggled with what could be called the "tyranny of distance," the fact that this region remains so far and so isolated from the other world regions. Most of the economies of these countries are based on exports, but these exports must be shipped, adding on to the cost of production. Anything that is not made locally must also be shipped in and some countries have promoted **import-substitution industries**, a strategy to replace foreign imports with domestic production of goods. Australia remains relatively unique among more developed countries in that its economy is based heavily on the export of commodities. Geography plays a key role in Australia's export-oriented economy. The country is relatively large and has a significant amount of natural resources. Its domestic population compared to its size and resource base is quite small, at only 24.6 million people as of 2017, so it is able to export the resources it doesn't need domestically. The country is the global leader in coal exports and has the second-largest diamond mine in the world.

Across the smaller islands of the Pacific, geography has also played a key role in economic development. These countries are often very small with only limited natural resources. They are also remote, only connected by long shipping routes to other world regions. Many rely on trade to Australia, New Zealand, China, Japan, and the United States for both export income and imported goods. Often, residents of the Pacific islands speak **Pidgin English**, a simplified form of English used by speakers of different languages for trade, in addition to their native tongue. Tourism has become a significant source of income for some countries of the Pacific, especially for the islands of Fiji. Despite some economic successes, these are also some of the world's most vulnerable countries in terms of global changes in climate and the coming decades could see significant changes in the human landscape as a result.

10.5 THE CHANGING LANDSCAPE OF OCEANIA

The human settlement of Oceania, from the earliest migrations to European colonization, has reshaped the physical landscape of this region. **Environmental degradation**, disturbances to resources like air, land, and water, is a serious concern as economic growth often comes at the expense of environmental sustainability. In Australia, for example, wide stretches of previously sparsely inhabited Outback have become grazing lands. In Papua New Guinea illegal logging has contributed to significant deforestation. Pollution from dairying in New Zealand has led to high levels of water pollution.

Invasive species have also had significant environmental impacts in a region that has been otherwise relatively isolated. Australia has strict quarantine laws in an attempt to limit damage from nonnative plants and animals. The country currently spends around $4 billion yearly in invasive weed management alone. Cats have been banned in parts of New Zealand where they pose a threat to local bird species. Rats brought by early European ships have presented a significant problem for many islands of the Pacific where they kill other plants and animals and also spread disease. Offshore, invasive fish and algae species have damaged fragile ocean ecosystems.

In addition to local pollution concerns, human settlement of other world regions has contributed to pollution in the Pacific Ocean. Worldwide, there are five main ocean gyres, large systems of rotating ocean currents (see **Figure 10.12**). In the northern Pacific Ocean, one gyre has very high concentrations of trash and plastics carried to the area by ocean currents. It has been termed the Great Pacific garbage patch. When you throw something "away" improperly, these gyres are really where "away" is. A water bottle improperly disposed of on the western coast of North America will make its way to the Great Pacific garbage patch in around six years.

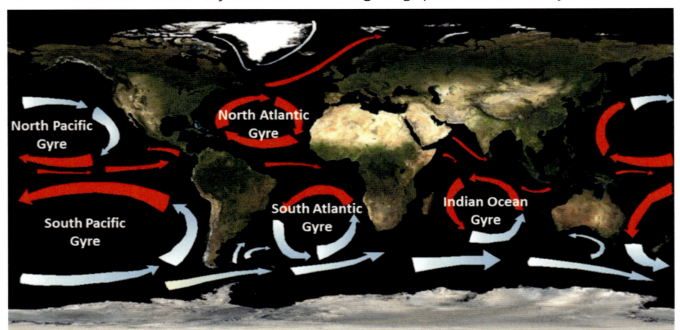

Figure 10.12: Ocean Gyres and Currents (Derivative work from original by Ingwik, Public Domain)

One issue with plastics is that they don't biodegrade, and instead keep breaking down into smaller pieces while still remaining plastic. Fish and other marine life eat these tiny bits of plastic, which can disrupt a number of biological systems. Some of these fish species are then consumed by humans. Because the Great Pacific garbage patch contains such small pieces of plastic, and most of the plastic is below the surface of the ocean, it is not easily visible with the naked eye and it is difficult to estimate its size. Some have theorized the patch is as big as or bigger than the US state of Texas, while others note that the idea of a "patch" of garbage is really a misnomer as there are concentrations of trash throughout the world's oceans.

Trash from other world regions also washes up along the shores of the Pacific islands. Kamilo Beach in Hawaii is the site of a significant amount of plastic that has washed ashore from the Great Pacific garbage patch, so much so that the area has been nicknamed "Plastic Beach" (see **Figure 10.13**). Though the shoreline looks sandy, 90 percent of it is actually bits of plastic and you would have to dig down at least one foot to reach grains of sand. Plastic trash litters many of the shorelines of the Pacific islands presenting a hazard for marine life and a management and cleanup challenge since debris often comes from thousands of miles away.

Figure 10.13: Plastic Debris on Kamilo Beach, Hawaii (Algalita.org, Public Domain)

It is changes in global climate, however, that pose the most severe environmental threat to Oceania. For many of the world's regions, changes in climate are viewed as hypothetical. Hurricanes might increase in intensity. The risk of fire might increase. Changes in bird migrations in Europe and North America to shifts in global fish stocks have already been linked to increases in global temperature but with little direct effect on the human populations of these regions. In

Oceania, though, small increases in temperature and ocean levels could have disastrous effects on already fragile ecosystems and economies.

The Great Barrier Reef is currently experiencing periods of **coral bleaching** due to increasing ocean temperatures. When waters get too warm, coral experience "stress" and expel the colorful, algae-like organisms that live within them. Mass coral bleaching has occurred several times since the late 1990s and is expected to become a regular occurrence as ocean temperatures continue to rise. Coral bleaching has also been documented in other reefs including ones in Hawaii.

Some of the leaders of the Pacific islands have been among the most vocal champions for global climate regulations. Speaking in 2015, the prime minister of Fiji, Frank Bainimarama did not mince words: "Unless the world acts decisively in the coming weeks to begin addressing the greatest challenge of our age, then the Pacific, as we know it, is doomed." Fiji has already experienced an increase in infectious diseases related to higher temperatures, record-breaking high tides, and has had to relocate citizens due to rising ocean levels.

Many of the effects of warming temperatures in the Pacific relate to changes that are first being documented in the Arctic. Scientists have consistently documented the melting of Arctic sea ice for the past decade (see Figure **10.14**). As the ice sheets melt, raising global sea levels, the surface becomes less reflective and absorbs more of the sun's rays, further accelerating warming and beginning a cycle that could be difficult to undo. Both the North and South Poles have experienced faster warming than the rest of the world, and some areas of the Arctic have seen an increase in temperature by 3 to 4 °C (5.4 to 7.2 °F).

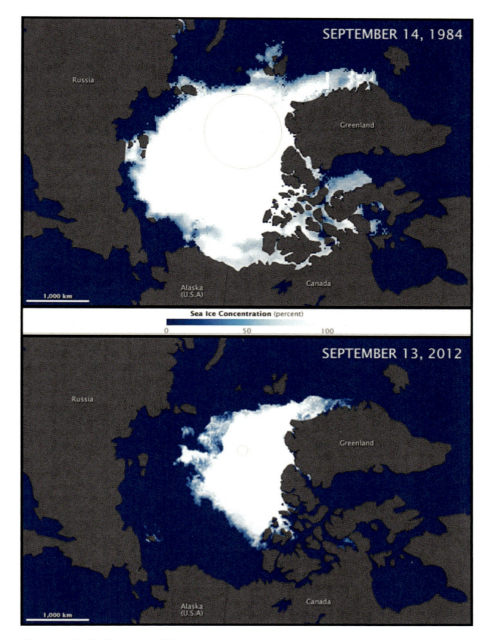

Figure 10.14: Minimum Extent of Arctic Sea Ice, 1984 and 2012 (Derivative work from original by Jesse Allen, NASA Earth Observatory, Public Domain)

The rising oceans in the Pacific and the concentrations of pollution found in this region are stark reminders of our interconnected world. Fossil fuel emissions from a car in the United States affect the amount of greenhouses gases in the atmosphere which in turn can increase global temperatures and melt ice in the Arctic. A plastic shopping bag discarded improperly in Japan makes its way out to sea where it breaks down and forms a coat of plastic sand on a Pacific island. But this same interconnectedness could perhaps be used to our advantage. Small, individual changes magnified across the global system could have profoundly positive effects. In many ways, the future of Oceania will be decided by the actions of global citizens and leaders.

Glossary

Aborigine
    the term for the indigenous people of Australia

absolute location
    references an exact point on Earth and commonly uses specific coordinates like latitude and longitude

acid rain
    a form of acidic precipitation caused by the emission of sulfur dioxide and nitrogen oxide from the burning of fossil fuels

African Union
    an interregional organization in Africa that seeks unity, integration, and sustainable development

agricultural density
    the ratio of the number of farmers to the area of land

al-Qaeda
    a group of militant Sunnis founded by Osama bin Laden

Altiplano
    a series of high elevation plains found in western South America

altitudinal zonation
    distinct agricultural and livestock zones resulting from changes in elevation

apartheid
    the ruling Dutch government's policy of racial separation in South Africa

aquifers
    an underground layer of permeable rock that holds groundwater

Arab Spring
    a wave of protests and revolutions in North Africa and Southwest Asia that began in Tunisia in 2010

archipelago
a chain of islands

arithmetic density
the number of people in an area divided by the size of the area

Asian Tigers
refers to Hong Kong, Singapore, South Korea, and Taiwan which have experienced rapid industrialization and economic development led by export-driven economies, low taxes, and free trade, sometimes also called the Four Asian Tigers

assimilation
when one cultural group adopts the language and customs of another group

Association of Southeast Asian Nations
an international organization that seeks to promote political security, economic growth, and social development among member countries in Southeast Asia, also known as ASEAN

atoll
a ring-shaped chain of coral islands surrounding a central lagoon

biodiversity
having a wide variety of species present in an area

biogeography
a branch of geography that explores the spatial distribution of the world's flora and fauna

Bolsheviks
a Marxist political party led by Vladimir Lenin that overthrew the interim government following the Russian Revolution and created the Union of Soviet Socialist Republics

boreal forest
a cold biome characterized by coniferous trees, also known as taiga

brain drain
refers to the emigration of highly skilled workers "draining" their home country of their knowledge and skills

Buddhism
religion that emerged from Hinduism and is based on the teachings of Siddhartha Gautama

buffer states
a country situated between two more powerful states

caste system
a form of hereditary social hierarchy found in Hinduism

central business district
    the central commercial and business area of a city, also known as the CBD

centrifugal forces
    forces that threaten national unity by dividing a state

centripetal forces
    forces that tend to unify people within a country

choke point
    a narrow passage to another region, such as a canal, valley, or bridge

Christianity
    a monotheistic religion based on the life and teachings of Jesus

climate change
    global changes in temperature and the patterns of weather over an extended period of time

Cold War
    a time of political and military tension primarily between the United States and the Soviet Union following World War II and lasting until the early 1990s with the fall of the Soviet Union

colonialism
    the control of a territory by another group

command economy
    an economic system where the production, prices of goods, and wages received by workers is set by the government

commodities
    raw materials or agricultural goods that are easily bought and sold

communal conflict
    violence between members of different communities

communism
    a social, political, and economic system that seeks communal ownership of the means of production

continental climate
    areas near the center of a continent that experience more extremes in temperature due to their location away from bodies of water

coral bleaching
    occurs when coral experience "stress" due to warm waters and expel the colorful, algae-like organisms that live within them

deindustrialization
    the process of shifting from a primarily manufacturing-based economy to service industries

demographic transition model
    a model that demonstrates the changes in birth rates, death rates, and population growth over time as a country develops, also referred to as the DTM

desertification
    the process of previously fertile land becoming desert

development
    economic, social, and institutional advancements

diffusion
    refers to the spreading of an idea, object, or feature from one place to another

distortion
    changes that occur in area, shape, distance, and/or direction when representing a spherical Earth on the flat surface of a map

domino theory
    refers to the fear that the fall of one country to communism would lead to the fall of other surrounding countries to communism

dual economy
    when plantations or commercial agriculture is practiced alongside traditional agricultural methods

Eastern Orthodox Church
    a branch of Christianity that split from Roman Catholicism in 1054 CE and includes a number of different denominations such as the Russian Orthodox and Greek Orthodox churches

economies of scale
    the savings in cost per unit that results from increasing production

El Niño
    the warming phase of a climate pattern found across the tropical Pacific Ocean region

endemic
    a disease found within a population in relatively steady numbers

entrepôt
    a French term meaning a commercial center of trade

environmental degradation
    the deterioration of resources like air, land, and water

epidemic
    a disease outbreak

ethnicity
    the identification of a group of people with a common language, ancestry, or cultural history

exclusive economic zone
    a 200 mile zone extending out from a country's coastline where it has exclusive control over any natural resources, also known as the EEZ

failed state
    when a government deteriorates to the point where it is no longer functional

federal state
    a political system characterized by regional governments or self-governing states

Fertile Crescent
    an early area of human civilization in North Africa and Southwest Asia surrounding the Tigris, Euphrates, and Nile rivers

floating population
    members of a population who reside in an area for a period of time but do not live there permanently

foreign direct investment
    the control of a business in one country by a company based in another country, also known as FDI

formal regions
    a region that shares at least one common characteristic, sometimes also called homogeneous regions

forward capital
    a capital that has been intentionally relocated, generally because of economic or strategic regions, and is often positioned on the edge of contested territory

fossil fuels
    nonrenewable sources of energy formed by the remains of decayed plants or animals

functional region
    a region united by a particular function, often economic, sometimes also called nodal regions

genocide
    the systematic elimination of a group of people

gentrification
   where increased property values resulting from the arrival of middle and upper class residents displace lower-income families and small businesses

geographic information science
   also referred to as geographic information systems, or GIS; a program that uses computers and satellite imagery to capture, store, manipulate, analyze, manage, and present spatial data

globalization
   the increasing interconnectedness and integration of the countries of the world resulting from advances in communication and transportation technology

Great Barrier Reef
   a massive underwater coral reef off the coast of northeastern Australia

Great Leap Forward
   a campaign begun in 1958 by the Communist Party of China that sought to reshape China's agrarian society into an industrial power

Green Revolution
   changes in agricultural technology and productivity beginning in India in the 1960s

gross domestic product
   the value of all the goods and services produced in a country in a given year

gross national income
   the value of all the goods and services produced in a country in a given year and the income received from overseas

Group of Seven
   a political forum of the world's leading industrialized countries including Canada, France, Germany, Italy, Japan, the United Kingdom, the United States, and the European Union, also known as the G7

growth pole
   a cluster within a region that has attracted economic development

hacienda
   a Spanish estate where a variety of crops are grown both for local and international markets

hajj
   a pilgrimage to Mecca that is expected for all physically and financially able Muslims to complete at least once in their lifetime

high islands
   islands that were formed from volcanoes and have relatively high relief

Hinduism
an ancient polytheistic religion that first developed in South Asia and is characterized by a belief in karma, dharma, and reincarnation

Horn of Africa
a protruding peninsula in East Africa that contains the countries of Djibouti, Eritrea, Ethiopia, and Somalia

humanism
a philosophy emphasizing the value of human beings and the use of reason in solving problems

import-substitution industry
a strategy to replace foreign imports with domestic production of goods

Industrial Revolution
the changes in manufacturing that began in the United Kingdom in the late 18th and early 19th centuries

informal sector
refers to the part of the economy where goods and services are bought and sold without being taxed or monitored by the government

insurgent state
a territory beyond the control of government forces

invasive species
a species of plant, animal, or fungus that is not native to an area but spreads rapidly

Iron Curtain
an imaginary dividing line between the Soviet Union and its satellite states who aligned with the Warsaw Pact, a collective defense treaty, and Western European countries allied through the North Atlantic Treaty Organization (NATO)

Islam
a monotheistic religion that emphasizes the belief in Muhammad as the last prophet

Islamism
a religious ideology characterized by a strict, literal interpretation of the Qur'an, conservative moral values, and the desire to establish Islamic values across the entire world.

isthmus
a narrow strip of land that connects the two large landmasses

Jainism
a religion emerged in India in the first century BCE and emphasizes ahimsa, nonviolence toward all living beings

jihadism
    a militant form of Islam that seeks to combat threats to the Muslim community

Judaism
    an ancient monotheistic religion founded in the Middle East that holds the Torah as its holiest religious text

Khmer Rouge
    a communist organization in Cambodia that opposed Westernization and believed in a return to an agrarian society

land alienation
    when land is taken from one group and claimed by another

latitude
    imaginary lines that run laterally, parallel to the equator, around the earth and measure distances north or south of the equator

liberation theology
    a form of Christianity that is blended with political activism and places a strong emphasis on social justice, poverty, and human rights

lingua franca
    a common language spoken between speakers of different languages

longitude
    imaginary lines that circle the earth and converge at the poles, measuring distances east and west of the equator

low islands
    islands that were formed mostly from coral and have relatively low elevations

Maori
    the term for the indigenous people of New Zealand

maquiladora
    a manufacturing plant that takes components of products and assembles them for export

maritime climate
    a climate zone that features cool summers and cool winters with few extremes in temperature or in rainfall

Marxism
    an analysis of social class and conflict based on the work of Karl Marx

megacity
    a metropolitan area with over 10 million people

megalopolis
    a region of overlapping metropolitan areas

Melanesia
    a region of islands to the northeast of Australia that include Papua New Guinea and Fiji

melting pot
    a metaphorical term referring to the mixing of cultural groups to create a more homogeneous national culture

mestizo
    a term referring to someone of mixed European and Ameridian descent

microfinance
    financial and investment services for individuals and small business who otherwise do not have access to traditional banking services

Micronesia
    a region of very small islands north of Melanesia and east of Polynesia

migration
    a move from one place to another intended to be permanent

monotheistic
    the belief in one god

monotreme
    an egg-laying mammal

monsoon
    a seasonal shift in winds that results in changes in precipitation

nation-state
    a sovereign political area that has a homogenous ethnic and cultural identity

nationalism
    the feeling of political unity within a territory

neocolonialism
    the practice of exerting economic rather than direct political control over territory

Neolithic Period
    also known as the New Stone Age, a time of key developments in early human technology, such as farming, the domestication of plants and animals, and the use of pottery

North American Free Trade Agreement
    an agreement established in 1994 with the goal of increasing economic cooperation between Canada, Mexico, and the United States, also referred to as NAFTA

offshore banking
financial services located outside a depositor's country of residence and offer increased privacy and little or no taxation

orographic precipitation
rainfall that results from a physical barrier forcing air masses to climb where they then cool, condense, and form precipitation

Outback
a remote area of extensive grassland pastures in central Australia

outsourcing
contracting out a portion of a business to another party, which might be located in a different country

Pacific Rim
refers to the countries that border the Pacific Ocean

Pacific Ring of Fire
an area of high tectonic activity along the Pacific Ocean basin

partition
the division of a territory into smaller units, as with the former British Empire in South Asia

permafrost
soil that is consistently below the freezing point of water (0°C or 32°F)

physiologic density
the number of people per unit of arable land

Pidgin English
a simplified form of English used by speakers of different languages for trade, in addition to their native tongue

plantation
an agricultural system designed to produce one or two crops primarily for export

plate tectonics
a theory that describes the movement of rigid, tectonic plates above a bed of molten, flowing material

Polynesia
a large, triangular region of Pacific islands stretching from New Zealand to Easter Island to the Hawaiian and Midway Islands

population pyramid
> a graphical representation of a population's age groups and composition of males and females

primate city
> a city that is the largest city in a country, is more than twice as large as the next largest city, and is representative of the national culture

qanat
> a system of irrigation first developed in modern-day Iran consisting of an underground tunnel used to extract groundwater from below mountains and transport it downhill

Qur'an
> the holiest book of Islam believed to contain the words of God as recited by Muhammad, also spelled Quran or Koran

rain shadow
> a region with dry conditions on the leeward side of a highland area

rate of natural increase
> the measure of a country's population growth calculated by subtracting its death rate from its birth rate, also referred to as RNI

refugees
> people who have been forced to leave their country

relative location
> the location of a place relative to other places

Russification
> a system of cultural assimilation in Russia where non-Russian groups give up their ethnic and linguistic identity and adopt the Russian culture and language

Sahel
> a transitional region in northern Africa connecting the dry Sahara Desert to the tropical regions of the south

scale
> the ratio between the distance between two locations on a map and the corresponding distance on Earth's surface

secularism
> the exclusion of religious ideologies from government or public activities

sex ratio
> the ratio of males to females in a population

shatter belt
an area of political instability that is caught between the interests of competing states

shifting cultivation
a system of agriculture where one area of land is farmed for a period of time and then abandoned until its fertility naturally restores

Sikhism
a monotheistic religion founded on the teachings of Guru Nanak that combines elements of both Hinduism and Islam

Slavs
an ethno-linguistic group located in Central and Eastern Europe that includes West Slavs (such as Poles, Czechs, and Slovaks), East Slavs (including Russians and Ukrainians), and South Slavs (namely Serbs, Croats, and others)

Small Island Developing States
small, coastal states with that face challenges related to sustainable development and have limited populations and natural resource bases, also known as SIDS

social stratification
a system of social categorization where people in a society have differing levels of social status

social welfare
a government policy where citizens pay a higher percentage of taxes to support universal healthcare, higher education, child care, and retirement programs

Special Economic Zones
a special region in China where more free-market oriented economics are allowed, also known as SEZ

squatter settlement
a housing area where residents do not own or pay rent and instead occupy otherwise unused land

state
an independent and sovereign political entity recognized by the international community

steppe
a biome characterized by treeless, grassland plains

structural adjustment programs
a set of required economic changes that accompany a loan made by the International Monetary Fund and the World Bank to country that is experiencing an economic crisis

subduction zone
an area where one tectonic plate is subducting, or moving below, another plate

subsistence farming
　where farmers grows food primarily to feed themselves and their families

sustainable agriculture
　a type of agriculture looks at farming's effect on the larger ecosystem and seeks to minimize negatively impacts on the ecosystem in the long-term

taiga
　a biome characterized by cold temperatures and coniferous forests

territorial waters
　an area extending 12 miles offshore that is considered sovereign territory of a state

theocracy
　a government ruled by religious authorities

total fertility rate
　the average number of children born to a woman during her childbearing years, also referred to as TFR

Trans-Siberian Railway
　an east-west rail line completed in 1919 that stretches across Russia, connecting Moscow in the west with Vladivosktok in the east

transition zone
　n area between two regions that is marked by a gradual spatial change

tribe
　a group of families united by a common ancestry and language

tsunami
　a high sea wave resulting from the displacement of water

tundra
　a biome characterized by very cold temperatures and limited tree growth

unitary state
　a political system characterized by a powerful central government

urban decentralization
　the spreading out of the population that resulted from suburbanization

urban sprawl
　the expansion of human settlements away from central cities and into low-density, car-dependent communities

urbanization
　the increased proportion of people living in urban areas

value added goods
a raw material that has been changed in a way that enhances its value

Wahhabism
a strict form of Sunni Islam that promotes ultraconservative Muslim values

Westernization
the process of adopting Western, particularly European and American, culture and values

World Trade Organization
an intergovernmental organization that collectively regulates international trade, also known as the WTO